Exposed
Structure
in Building
Design

Exposed Structure in Building Design

Charles H. Thornton
Richard L. Tomasetti
Janice L. Tuchman
Leonard M. Joseph

McGraw-Hill, Inc.

New York   San Francisco   Washington, D.C.   Auckland
Bogotá   Caracas   Lisbon   London   Madrid   Mexico City
Milan   Montreal   New Delhi   San Juan
Singapore   Sydney   Tokyo   Toronto

Library of Congress Cataloging-in-Publication Data

Exposed structure in building design / Charles H. Thornton...   [et al.].
     p.   cm.
   Includes bibliographical references and index.
   ISBN 0-07-064538-8
   1. Structural engineering. 2. Buildings. I. Thornton, Charles H.
TA636.E93   1993
720'.47—dc20   92-27933
                                                               CIP

1 2 3 4 5 6 7 8 9 0 DOHDOH 9 7 6 5 4 3 2

ISBN   0-07-064538-8

The sponsoring editor for this book was Joel Stein, the editing supervisor was Joseph Bertuna, the designer was Susan Maksuta, and the production supervisor was Pamela A. Pelton. It was set in Century Expanded on a Macintosh system.

Printed and bound by R. R. Donnelley & Sons Company.

Contents

About the Authors

Charles H. Thornton, Ph.D., P.E., is chairman and principal of Thornton-Tomasetti Engineers, an internationally renowned structural design firm based in New York City. He is a recognized expert in structural design, including the design of long-span and high-rise buildings, and in the behavior of materials exposed to low temperatures. Richard L. Tomasetti, P.E., is president and principal of Thornton-Tomasetti Engineers. He is a noted authority on structural performance, materials applications, and the design of unique structural systems and tall buildings. Janice L. Tuchman is familiar with the contemporary building industry as managing editor of McGraw-Hill's weekly news magazine, *Engineering News Record*. Leonard M. Joseph, P.E., is a senior associate at Thornton-Tomasetti Engineers with many years of design experience for a wide variety of structural materials and systems.

Preface

All buildings require structural frames for support; *exposed* structure can offer additional benefits. Exposing or expressing structure can make good economic sense, and highlighting structure in selected areas can create exciting spaces and a distinctive building identity. A successful design with exposed or expressed structure, however, requires close coordination among the various members of the design team.

Many books discuss overall building design issues, and other sources cover the technical issues of structural design, but few study their interaction. In this book, we explore the boundary where the two areas most visibly meet—exposed structure. This book is intended for use by people involved in buildings—architects, engineers, owners, contractors, and also by students and interested lay persons. A bibliography provides suggestions for further readings.

The book addresses structure in building design through two approaches: first, in a series of conversations, leading architects and engineers explain their methods of design, discuss the ways they handle design team interaction, and share their individual experiences. We hope they will stimulate others to think of building design in new ways and will also offer moral support to designers. Yes, integrating exposed structure involves a lot of give-and-take, but these professionals show that it can be done with fine results.

Second, in the technical chapters, we highlight the items designers should consider when exposing structure. This information is not intended to be exhaustive, since other available sources cover individual topics in great detail. Instead, we introduce the basic structural principles that apply to a topic, and then describe how exposing structure relates to or is affected by those principles. We show that the design of exposed structure is neither child's play nor impossibly arcane, but is practical, given suitable attention to detail.

This book involved the effort and input of many people. First, of course, are the architects and engineers interviewed. Their contribution is the heart of this book, and we thank them all: Edward Larrabee Barnes, Eli Cohen, Vincent DeSimone, Alfredo De Vido, Eugene J. Fasullo, James Ingo Freed, Hal Iyengar, John M. Y. Lee, William LeMessurier, Matthys Levy, Walter P. Moore, Gyo Obata, Cesar Pelli, Peter Rice, Leslie E. Robertson, Kevin Roche, Richard

Rogers, Bernard Tschumi, and Loring A. Wyllie. In addition, helpful information about concrete came from Jeremy Wood and Fruma Narov; Peter Pearce offered insights on long spans, and I. Paul Lew was a great help with parking structures.

Preparing the manuscript involved a large effort. It could never have been completed without our families' forbearance during writing sessions, and Nadine Post's inspiration, Howard Stussman's encouragement, Barbara Solomon's research, Aranza Winkelman's graphics assistance, and especially Mary Murray's patient transcribing, typing, coordinating, and all-around organizing. We also wish to thank anyone else who inadvertantly is not mentioned here. Thank you one and all.

Charles H. Thornton
Richard L. Tomasetti
Janice L. Tuchman
Leonard M. Joseph

1

To Expose...
Or Not?

The bold Xs of the John Hancock building have become part of the identity of Chicago. They are an architectural statement: This is Big John—strong, tall, and proud. But the steel Xs are also the cross bracing of the 100-story building, stabilizing it against enormous lateral forces (see Fig. 3.13). Across the city, an office building called Onterie Center repeats the giant Xs, this time in painted concrete (see Fig. 3.14). But whether concrete or steel, the combination is at once an architectural element and a structural system—exposed structure.

Exposed structure is found all over the world. In Hong Kong (Fig. 1.1) the Hongkong and Shanghai Bank (at left) and the Bank of China (at right) show two very visible, very different approaches. Exposed concrete frames also occur throughout the city.

Exposed structure is by no means found only in office buildings or executed only in steel or concrete. At the United Airlines Terminal at O'Hare International Airport, exposed steel arches gracefully vault above the concourses (see Fig. 5.13). At Pontiac Stadium in Michigan, the roof is an expanse of air-supported, cable-restrained, Teflon-coated fiberglass fabric. At the Georgia Dome, Teflon-coated fiberglass fabric is supported by cables and posts in a "tensegrity dome" configuration (see Figs. 3.22 and 3.23).

In some exposed structures, the structural material carrying load is actually in contact with the elements and is subjected to changes from temperature and from weathering. In other cases, painting or coating may protect the structure from weathering, but still leave it subject to temperature-induced expansion and contraction. Sometimes a structural system is on display, either inside or outside a building, but is covered or encased in another material to insulate it or to offer better durability or aesthetics. In these cases the structural system is expressed but not literally exposed. An example is the granite-clad diagonals of the Wang Building in New York City (Fig. 1.2). Because the expression of structure often has characteristics in common with exposed structures, it will be discussed in this book as a variation on the theme.

Exposing structure can be a cost-effective approach to building

Figure 1.1 This view of Hong Kong clearly shows two very different forms of structural expression—Hongkong and Shanghai Bank at left and Bank of China at right. Surrounding buildings also have exposed frames.
Hongkong and Shanghai Bank; Architect: *Foster Associates*; Engineer: *Ove Arup & Partners*. **Bank of China**; Architect: *I. M. Pei and Partners*; Engineer: *Leslie E. Robertson Associates*. (Photo: © *Paul Warchol*.)

structures large and small, spanning long distances, creating large interior spaces, and handling crowds. The technique has been used in high-rise office buildings, in small industrial plants, in exhibition halls, atriums, and stadiums. The economic advantage of having a single material serve a dual purpose is obvious.

But the technique is not as simple as it seems. Behind the advantages are hidden pitfalls. The savings may not materialize if excessive fastidiousness about how an exposed structure will be perceived leads to special details and treatments that drive up its cost.

Figure 1.2 The Wang Building has its cross-braced structure expressed, if not actually exposed.
780 Third Avenue, New York, N.Y.; Architect and Engineer: *Skidmore, Owings and Merrill.* (Photo: *Len Joseph.*)

Plaster and joint compound can be smoothed more easily than steel and concrete!

Exposing the structure is potentially one of the most dramatic gestures in the repertoire of the architect. However, while on one building it can take courage to expose the structure, on another building it can be gutsy to hide it. The exposed structure of the Hongkong and Shanghai Bank is viewed as the heroic, tour-de-force solution. The most casual observer can see that the floors are hung from towers (Fig. 1.3), and that the towers are stabilized by cross bracing and "ladder rungs" (Fig. 1.4). But concealing the support system for the cantilevered corner over the entry of the IBM building in New York City is an equally bold statement, deriving its dramatic tension from the lack of visible means of support (Fig. 1.5). A notched corner also signals the entry to 535 Madison Avenue, but the addition of a corner column creates a more secure feeling (Fig. 1.6). At Citicorp, setting the building on stilts also creates great drama, but the symmetrical midface legs induce a sense of stability (see Fig. 3.17).

Most mainstream architecture neither exposes nor expresses structure, seeking the safety of conventional architectural finishes. Achieving an aesthetically acceptable and universally acclaimed building while working within the constraints of a structural system can require more care and planning than covering up structure with purely architectural materials. Gone is the space between cladding and structure—the area where architects and mechanical and electrical engineers have always threaded electrical conduits, ducts, and piping without exposing them to view. For example, roof drainage for a totally exposed roof structure can pose a challenge since it is quite difficult to make drains aesthetically appealing. How to make it clear that this is a drain line and not a column? Or should it be perceived as a column? At the classic suspended roof of Dulles Airport, for example, the main drain is given an appropriate "aerodynamic" treatment (Fig. 1.7).

Designers who expose structure must also consider carefully the practicality of maintaining a uniform appearance of materials. Most structural products are not quality-controlled for appearance or manufactured to have uniform color or texture. Yet most people expect a wall to have a uniform color, whether it is architectural concrete or an expensive finish brick. By providing viewers with something larger to focus on, rustication strips, control joints, and other patterns can reduce the need to match fine textures and color. Joint patterns can also relate to module spacing and provide a

Figure 1.3 The path of gravity loads—hung floors, hangers, sloping tension arms, and load-bearing towers—is clearly expressed on this facade. Interrupting the hangers emphasizes their function.
Hongkong and Shanghai Bank, Hong Kong; Architect: *Foster Associates*; Engineer: *Ove Arup & Partners*. (Photo: *Len Joseph*.)

Figure 1.4 Cross bracing and "ladder rungs" visible within the lobby remind visitors of the strength of this structural system.
Hongkong and Shanghai Bank, Hong Kong; Architect: *Foster Associates*; Engineer: *Ove Arup & Partners.* (Photo: *Len Joseph.*)

Figure 1.5 This cantilevered corner entrance hints at the power of structure hidden behind the blank facade above.
IBM Building, New York, N.Y.; Architect: *Edward Larrabee Barnes/John M. Y. Lee P.C.*; Engineers: *The Office of James Ruderman and LeMessurier Associates.* (Photo: *Len Joseph.*)

Figure 1.6 A column clearly supports this corner entrance, carved out to open views to a small park.
535 Madison Avenue, New York, N.Y.; Architect: *Edward Larrabee Barnes/John M. Y. Lee P.C.*; Engineer: *The Office of Irwin G. Cantor.* (Photo: *Len Joseph.*)

Figure 1.7 The sweep of this suspended roof is clearly visible, as is the logic behind the aeronautically styled main roof drain.
Dulles International Airport Main Terminal, Washington, D.C.; Architect: *Eero Saarinen*; Engineer: *Ammann and Whitney*. (Photo: *Len Joseph*.)

Figure 1.8 An exception to the rule of interrelated form and function, these precast, posttensioned shells cover separate acoustically designed performance spaces which stand within.
Sydney Opera House, Sydney, Australia; Architect (stages 1 and 2): *Jørn Utzon*; Engineer: *Ove Arup & Partners*. (Photo: *Daniel Cuoco*.)

desirable sense of proportion (see Figs. 4.2 and 6.1).

Another issue to consider when exposing structure is resolving the potentially conflicting demands of planning modules for architectural and structural purposes. All structures follow some module—generally set by the planning of the interior and its relationship to the exterior. When the structural system is exposed, the structural module is superimposed on the architectural module. As a result, controlling the scale and texture of the building becomes extremely important. Setting a module which accommodates both architecture and structure accomplishes this task. Of course, every rule has an exception. The Sydney Opera House "sails" are exposed structural elements of precast, posttensioned concrete, visible inside and out (Fig. 1.8). For acoustic reasons the performance spaces are of separate construction, with little relationship to the marvelous enclosure. Clearly this approach is feasible only for a "signature" project.

Exposed and expressed structures also require consistent patterns of perception. For example, when a long beam spanning 50 feet frames into a short girder spanning 10 feet, the long beam would usually be the deeper of the two members. If concealed above a ceiling, the discrepancy in member sizes would not be important. But the architect of an exposed structure, attempting to anticipate public perceptions, may feel that the girder should appear to be as strong as the beam—and that may mean it should be as deep or deeper.

A similar condition occurs in high-rise buildings with lobby-level columns that do not participate in the wind system. If they do not resist heavy wind loads, the columns may have no structural reason to be very large in plan. But regardless of structural need, the architect may want them to look substantial to give the building the appearance of solidity. At the Georgetown Plaza in New York City, for example, where steel columns support a concrete transfer system, the architect actually enclosed the steel in a round masonry drum to make the building appear more solid. Conversely, at the AT&T Building, where wind is taken by internal elevator cores, the columns were intentionally expressed as tall, slender elements for a loggia effect (Fig. 1.9). When exposing or expressing structure, the architect will develop and suggest relative proportions. It is essential that the structural engineer assist the architect in establishing workable guidelines, and then stay within those guidelines to achieve an acceptable solution.

Builders have seen beauty and utility in exposed structure since prehistoric times—what else are tepees, huts, or, on a grander

Figure 1.9 Concealing the wind bracing system within the cores of this steel-framed structure permits the use of tall, slender stone-clad piers for this loggia.
AT&T Building, New York, N.Y.; Architect: *Philip Johnson—John Burgee Architects*; Engineer: *Leslie E. Robertson Associates*. (Photo: *Len Joseph.*)

---

scale, Stonehenge? With the Egyptians came the beginning of the architect's love affair with the column. The hypostyle hall at the Great Temple of Ammon at Karnak, built around 1300 B.C., derives its visual strength from rows of columns. And the pyramids are, of course, among the most imposing works of exposed structure of all time.

In the Parthenon and other temples built between 500 and 400 B.C., Greek architects refined the idea of the column as both structure and architecture. Roman architects used columns both as exposed structure and as decoration—to decorate the piers between arches at the Circus Maximus in Rome, for example. But they also added not just the arch, but the vault and the dome to the repertoire of exposed structural forms. The Pantheon, built in Rome in about 120 A.D., for example, has a tiered rotunda topped by a hemispherical dome 142 feet in diameter.

Medieval times may have represented a stagnant period in the intellectual life of Europe, but they did bring advances in exposed structure. The English raised the construction of open timber roofs to an art form, the most elegant example being the intricate hammer-beam type built in the 15th century. At the same time the French were building stone vaults braced by flying buttresses.

Exposed columns, arches, vaults, and domes continued to be popular through the Italian and French renaissance and on into the 19th century. At that point, the use of exposed cast iron made its appearance, most notably in bridges, railway stations, and industrial structures. For the international exhibition of 1889, architect F. Dutert and engineer Contamin designed the Halle des Machines— 375 feet wide, 1400 feet long, and 150 feet high—supported by four steel arches. The renowned engineer-architect Gustave Eiffel designed his 984-foot-high tower for the same exhibition.

Toward the end of the 19th century, a new material, reinforced concrete, was coming into its own and also found applications in exposed settings. Designers, too, began to free themselves from the classic forms. In 1916 came the astonishing airship hangar by the engineer Freyssinet, a ribbed thin-skin parabolic concrete vault. And in 1956, the Palazzetto dello Sport in Rome by Pier Luigi Nervi and Annibale Vitellozzi further developed this approach. A shallow shell concrete dome, 200 feet in diameter, is supported around the base by Y-shaped flying buttresses.

Current examples of exposed and expressed structure abound and cover a wide range. In some building types, such as parking decks, structure is exposed as a matter of function. In others, structure could be concealed, but is exposed as a matter of economics.

This is the case for many stadiums and arenas, such as the Georgia Dome stadium. A third category of building has structural systems developed for efficiency, and then exposed or expressed as an architectural design decision. "Big John," Onterie Center, and Bank of China (see Fig. 3.32) are examples. In a class by themselves are those buildings where the structure is conceived, designed, and expressed as part of the architecture. As "live-in sculpture" they create particularly striking images—the Eiffel Tower, Sydney Opera House, United Airlines Terminal, Hongkong and Shanghai Bank, Centre Pompidou—but they also require the highest degree of coordination between architects, engineers, contractors, fabricators, and erectors (see Figs. 1.8, 5.13, 1.4, 2.30).

In this book we explore the decision-making process behind exposed structures from the points of view of leading designers and highlight the special characteristics of exposed structure which should be considered during the design process. If, along the way, you are impressed by the broad range, functionality, and beauty of these buildings—that was our intention too!

2 *Conversations With Architects*

INTRODUCTION

Interviewing architects for this book has been an eye-opening experience. We began the project expecting to find a degree of unanimity, or at least common ground, among the current leaders of building design. After all, we "knew" that design styles—neoclassical, romantic, international, modern, postmodern, deconstructivist—have distinct periods, and that architects at the forefront would be pretty much in synch, right? Well...it didn't work that way. Designers who are accomplished and respected are also secure enough to maintain their own visions of what architecture should be, and they articulate their visions here without hesitation. Listen for the distinctive "voice" of each architect, and for the unique perspective each has on his own work and on the work of others.

EDWARD LARRABEE BARNES AND JOHN M. Y. LEE

The authors sought out New York City-based architect Edward Larrabee Barnes and his partner John M. Y. Lee to discuss the dramatic 60-foot cantilever over the corner entrance of the IBM corporate headquarters at 590 Madison Avenue. The bold structural statement is in strong contrast to another building the firm designed only blocks away, at 535 Madison Avenue. Here Barnes and Lee created a similar open corner entryway, but this time they supported the overhang with a column. What led the firm to make such different choices? Their decisions were well thought out. Interestingly, the architects' point of view is quite different from the layperson's. Barnes and Lee feel that the IBM cantilever is actually more expressive of the building's true structure than the supported overhang at the 535 Madison entrance.

THORNTON: My perception of your firm's work is that there's been a tendency to not express structure. One of my favorite buildings you designed is the IBM World Trade building up in North

Tarrytown with the pond and the geese. It fits into the site so well. But it does not express the structure.

LEE: Is that true? I think it does express the structure.

THORNTON: Well, what I like about that building is that the building itself is unobtrusive. You drive by and it just sort of fits in. Are you saying that if you look closely, you can tell what the structural approach is?

LEE: Absolutely! The floor, it's cantilevering. There's a 10-foot cantilever, so the floor is expressed and the columns are set back (Fig. 2.1). But in the back where we don't have the cantilever there are column modules, and all that is expressed.

You're correct in the sense that our first priority is of volume massing. The first thing we do is make block models. We are pretty conscious of giving people a sense of structure, even though we don't literally expose the structure.

THORNTON: The interesting conclusion I draw when I sit down and talk with architects is that my perception of whether or not you expose structure is generally different than your perception. The position of whether or not you are expressing the structure is also relative to who's designing it and who's perceiving it.

LEE: That's why I say you have to have two categories. One is what you see almost instinctively. There is a sense of structure that gives you the scale in relation to the structure, versus the purely exhibitionist approach to structure.

THORNTON: The English architects Norman Foster and Richard Rogers have become exhibitionists relative to exposed structure. United Airlines that we did with Helmut Jahn is also structural exhibitionism (see Fig. 5.13).

LEE: But in that context I think it depends on the building type. If we were given a chance to do an exhibition hall, a coliseum, or something like that, I am sure we would design a building in which the structure becomes quite prominent. If we do an office building, where enclosure of interior volume is important, then it's totally different.

THORNTON: If you go back 20 years and look at a number of classroom structures, there was a lot of concrete and exposed waffle construction; strong structural statements in a lot of schools. Today we seem to be going back to the more traditional brick and stone facades with less use of exposed structure.

LEE: But I think we do both. I think it really depends on the context of the buildings around the area. Maybe there's already a lot of so - called brick boxes in the area—then you try to maintain the context.

If we had a chance to do the Citicorp building, for example,

Figure 2.1 Structure is expressed here by setting columns back and letting the floors appear to float.
IBM World Trade Headquarters, Mt. Pleasant, N.Y.; Architect: *Edward Larrabee Barnes/John M. Y. Lee P. C.*; Engineer: *Severud, Perrone, Sturm, Bandel.* (Photo: *Joseph Molitor.*)

using the same principal structural system, we would somehow find a way to express the diagonal members because they are the main reason for the building to be. But Citicorp's architect obviously didn't try to do that (see Figs. 3.16 and 3.17).

THORNTON: It's easier to express it than to expose it because then you can keep the curtain wall intact. Are you familiar with Broadgate in London? It's an SOM design where they have an arch support over railroad tracks. They literally took steel arches and supported the whole building on them. The arches are 4 or 5 feet outside the glass line. And the steel is not only expressed, it's exposed (see Fig. 3.15). Kevin Roche's Knights of Columbus in New Haven is similar (Fig. 2.24).

LEE: Well I'm not saying whether we would do that in the sense of exhibiting the structure as such. But we certainly would try to do something to show that the diagonals are a very important thing to the building.

THORNTON: One of the questions I want to ask Ed is why there's no attempt to express whatever is holding up the cantilever on the IBM building (Fig. 1.5). Do you think that when people look at this building, they get afraid or concerned?

LEE: Well obviously a lot of people wrote about that, including Paul Goldberger. He mentioned that it's scary to him, the building, the whole volume with no apparent support. Evidently, that's how he feels. But I think if you ask average people how they think the corner is held up, they would say it's cantilevering.

BARNES: (*Entering*) Sorry I'm late....

THORNTON: One of the things John and I are talking about is your design for the IBM building versus your design for 535 Madison Avenue. Some people have said to me, "What's holding the IBM corner up?" For 535 what's holding it up is obvious (Fig. 1.6). John suggested that the 535 column was added because of economics.

BARNES: In addition to economic reasons, we wanted to mark the corner of the building to make it distinct from the adjoining pocket park, and somehow reinforce the street line.

THORNTON: You had the option to put in a corner column at IBM, but you made the decision not to. You also had an option to express how IBM was being supported and you didn't. What thought process went on in your mind when you were doing that?

BARNES: I think the question of whether to have a column or not in the first place is a result of the plan. The IBM building has a clear diagonal approach from the corner of 57th Street and Madison Avenue. It would have been all wrong to block this vista with a column.

As to the second part of your question, the question of expressing the cantilever structure, I feel that the broad band of stone extending out over the entrance *does* express structure— that this is a deep cantilever truss. (It is not necessary to be exhibitionist, to break through the skin with a structural exoskeleton.)

LEE: If I may add, there is a truss over the opening on 535, too.

BARNES: At 535 we are consciously keeping the rhythm of the windows, no matter what is happening behind the facade. The truss crosses behind the windows. So 535 is actually less expressive of the structure than IBM. On IBM you have a deep solid band which architects can read (Fig. 2.2). It is a place behind which you can have diagonal bracing and diagonal members. So while you might think IBM isn't as expressive, it is really more expressive than 535 Madison.

THORNTON: You and I know, as well as other architects and engineers, that the wide band means there is something structural back there. However, the lay person walking down the street would look up and not recognize this. I've heard people say that the corner without a column is scary.

Take the project we did together, 599 Lexington. We know that over the lobby opening there is a column that comes down and stops, a "shear link" column. Remember when we added those? There's an expression of structure there too. We didn't want to bring the columns all the way down, so we added a "shear link" column which participates in the wind system but doesn't carry gravity loads. It is expressed.

BARNES: What I am trying to say is that IBM does express the structure where 535 doesn't. While 535 may look more honest because of the column in the corner, it does not express the truss above the opening. The choice not to see the truss on 535 was a very conscious one. It has to do with the abstract metric quality of the architecture. The floor isn't expressed, it is completely camouflaged, so the same texture or module runs right through. That's a geometric decision, and it doesn't have anything to do with structure.

LEE: Also, I think when you either put the column in or take it out, the reason has nothing to do with structure or nonstructure. In the case of IBM this is the busy corner of the building where most people congregate.

BARNES: The column at 535 is really defining the park space, the rectangular plan of the park. You need something to mark that space of the street surface. Notice that the cutout on the bottom and the cutout on the top of the building are the same (Fig. 2.3). If this were simply an abstract object, you probably wouldn't put the column in. But the minute you start putting this on a site where it has to orient to the street and the park and so on, you need that definition of the two spaces at the lower level. So there's a choice here, not of subduing the structure but of reapportioning it. The reasons are site reasons.

IBM is a much simpler volume. There is a big truss floor. And all you're saying is, what if it had been an exoskeleton as opposed to something behind a skin? I don't know. There are some animals, crustaceans, and things that have exoskeletons. I think you could certainly rationalize those points of view.

Figure 2.2 This view clearly shows how the blank facade creates an impression of cantilever and counter-weight. It also shows the exposed framing of the greenhouse behind the tower.
IBM Building, New York, N.Y.; Architect: *Edward Larrabee Barnes/John M. Y. Lee P.C.*; Engineers: *The Office of James Ruderman and LeMessurier Associates.* (Photo: *Cervin Robinson.*)

Chapter Two

Figure 2.3 The corner column here serves to define the park, while permitting matching chamfers at base and top.
535 Madison Avenue, New York, N.Y.; Architect: *Edward Larrabee Barnes/John M. Y. Lee P.C.;* Engineer: *The Office of Irwin G. Cantor.* (Photo: © *Thorney Lieberman.*)

THORNTON: Just before you walked in we talked a bit about Citicorp and John said that if he had been designing Citicorp with the chevron scheme, he would have expressed it. Do you feel the same way?

BARNES: Yes, I think that in a big way Citicorp is masking too much, too many important things. Somehow I think that, weighing all these factors, a one-floor truss over the entrance is not that big a deal on 535, but on Citicorp the whole structural system of the building is masked with windows, all the way up the building.

Now if you look at 599 Lexington, our building, that was a case where we worked with diagonals (in plan). There the skin is set out from the structure a certain amount. The structure is not distorted; it is on module, and the skin is pulled out and pure. The impure thing on the skin, which we would have liked to have done something about, is that when you slice 45 degrees across a square (for the entrance and the upper setbacks) your window module should be 1.414 times the other module (Fig. 2.4a). That doesn't get anywhere with a real-estate man. We had to go to a 5-foot module north and south, east and west. But going northeast we had to get as close as we could to a 5-foot module. Ideally, if that building had been totally honest, then we would have liked that module changed to 1.414 times the 5-foot dimension.

THORNTON: When you start deviating from squares and rectangles, you run into some of these conflicts. I happen to feel that the exterior skin on 599 is one of the nicest that I've seen in a long time. It's wonderful, rich, deep, really great.

BARNES: But the skin of 599 is where the structure is not expressed. If you look at that skin carefully, you will see what appears to be a window is really a spandrel with the floor at the back. What appears to be a beam directly under the window is concealing a convector (Fig. 2.4b). That way the module comes out at the top of the building with metal instead of glass. Nobody picks it up, but if you really were expressing the cage, then you would expect the metal to be on the floor lines.

LEE: I think it's really a spectrum. If you don't pay attention to structure at all, you camouflage everything, it's just pictorial scenery. A lot of architects do that. Then sometimes you get a sense of structure, like from Mies and that kind of architect. You don't exactly see the structure, nuts and bolts, and all that, but you always sense it in the scale and the whole composition. Then you have a total exhibitionist structure, such as from Richard Rogers and Norman Foster and other people. But after saying that, I also say, it depends on the type of project. If we were doing a coliseum or exhibition hall, we probably would express the structure; all architects would do that because it relates to the scale of the space.

BARNES: But I think that total expression, like the Rogers kind of expression, which takes the bones of the building and puts them outside the skin so to speak, I think people have pretty well stopped doing that, haven't they?

THORNTON: It's very expensive. When we talked with Richard Rogers about his style of architecture for Lloyd's of London and Centre Pompidou, he didn't think you could find anybody in the United States who would be willing to foot the bill for that type of project (see Figs. 2.30 and 2.31). He is a realist; he knows that his architecture is expensive.

BARNES: I don't think we feel the same pressure on summer camp structures we've done, little buildings, even screened-in buildings where you don't have glass. In these buildings you can see the structure expressed with laminated wood construction. Fresh Air Fund camp buildings have rectangular bays; there's a lower beam and an upper beam and you express the nature of the building with post and lintel construction—beams and cross beams in two planes. The whole thing has a grid based on the spanning of the structure.

THORNTON: These are wooden structures?

BARNES: Yes, wood-laminated structures.

In Chinese and Japanese or other oriental buildings, when you get to the corner, it seems to me there is a lot of funny business. You're hard put to know what's happening, which beams are truly cantilevering and which are applied; with post and lintel there is an upper beam and a lower beam at the corner. It is quite clear how the two-way cantilevers are supported. At the camp, our office was doing about as simple a post and lintel construction as one can do, and expressing it.

THORNTON: What about projects where expressed or exposed structure is done in concrete? It seems to us that, in the United States, if you try to expose concrete structures internally or externally, the cost of formwork goes up far faster than, let's say, a hung ceiling, and so we always are fighting a battle. I find that we don't do many exposed concrete structures anymore because they come in at $800 a cubic yard instead of $300 a cubic yard.

BARNES: We have exposed concrete in a library we are doing in the midwest, just simple beam construction, and the concrete work is so bad. The contractor has to chip 2 inches off the face of some

Figure 2.4 (*a*) Although this building steps back three times, the skin was set to permit the structure to follow its preferred module. (*b*) The detailing of this facade creates the impression of a tall window to conceal the floor behind.
599 Lexington Avenue, New York, N.Y.; Architect: *Edward Larrabee Barnes/John M. Y. Lee P.C.*; Engineer: *Thornton-Tomasetti Engineers.* (Photos: © *Steve Rosenthal.*)

(a)

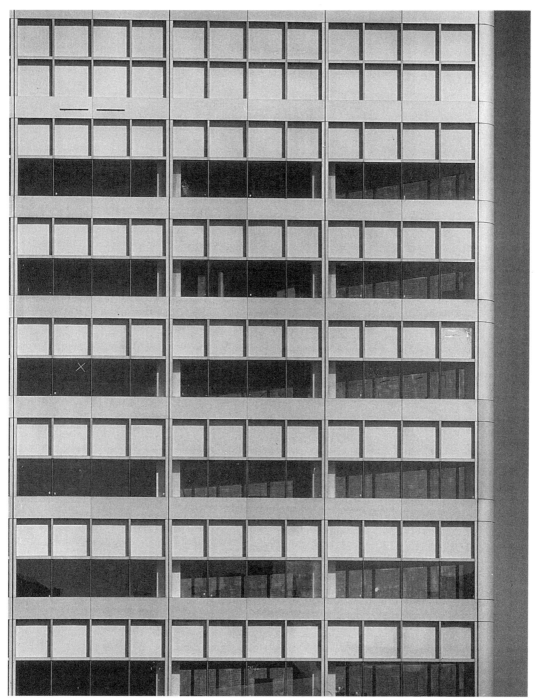

(b)

of the beams, they bulge; I don't know how we're going to make it look like what we want. It's a case which emphasizes what you are saying. If you're going to do concrete, you really have to jack up the specs, and then it gets expensive.

ALFREDO DE VIDO

Alfredo De Vido is a New York City-based architect who has used exposed structure on a variety of buildings, such as the original Wolf Trap Farm Park Performing Arts Center in Vienna, Virginia; Robin Hood Dell in Philadelphia; and the Snow King Inn in Jackson Hole, Wyoming. He has been particularly fascinated by exposed wood structures and has used them widely in office buildings and homes.

THORNTON: Is there really an economy in exposing structure, particularly for smaller buildings?

DE VIDO: If you just expose structure and then forget the finish, yes, it will be more economical. But you can't really do that, because if you're going to expose the structure, the key thing is to design the connections. You can't just hammer nails in and leave the beam dimpled. Care is necessary, or it will look bad. Frank Gehry has the reputation of just letting it all hang out. But if you look at his work, he takes great pains to make neat joinery. Exposed concrete is more economical if you do it conventionally. However, to get a superior finish, it is not—chances are it's going to be more expensive.

One of the problems of exposed structure is advising the trades of your intent. They make the structure sound, but don't have it programmed in their minds to use aesthetic care because the structure is usually covered up.

THORNTON: Is economics usually the driving force for you or aesthetics?

DE VIDO: At Wolf Trap there was a big roof with long spans entirely in wood, and other secondary "broken-back" beams that served to provide acoustically reflective panels. There it was economical because additional finish required would have added cost without aesthetics or acoustical dividends. However, the pattern of the beams then became important since they were exposed (Fig. 2.5). In the Robin Hood Dell auditorium there was a similar problem with the long-span steel joists used. To achieve an orderly pattern, we varied certain joist spacings, at very little cost (Figs. 2.6 and 2.7). But on the credit side, we saved huge amounts on the finish by exposing the structure.

In a concert hall the acoustics are extremely important. Most

Figure 2.5 This all-wood concert shell shows glued laminated girders, stiffened by tension rods lifting against steel pipe posts.
Wolf Trap Farm Park, Vienna, Va.; Architects: *MacFadyen/Knowles/De Vido*; Engineer: *Lev Zetlin Associates Inc./Thornton-Tomasetti*. (Photo: *Bill Maris*.)

finish materials are light and absorb low-frequency sounds. It is necessary to get reflective material that's dense enough, and wood decking is very good for that purpose. It is also aesthetically pleasing, with visual liveliness; difficult to achieve with plaster or sheet rock.

At an existing concert hall in Detroit they were losing low-frequency sound and lacked volume. We looked into improving the acoustics by stripping the interior out and exposing the structure, which was a crude concrete frame infilled with block.

Figure 2.6 This open-air concert hall uses steel framing in a simple geometry. **Robin Hood Dell, Philadelphia, Pa.**; Architects: *MacFadyen/De Vido*; Engineer: *Lev Zetlin Associates Inc./Thornton-Tomasetti.* (Photo: *Lawrence S. Williams.*)

No one had ever assumed for a second that it would be exposed. But we came up with the idea of providing an interior transparent lattice that would have a surface brightness achieved with lights which would stop the eye from the exposed crudely built structure. It would have turned the trick acoustically.

THORNTON: But at the Snow King Inn in Jackson Hole you had all kinds of options. You didn't have to expose the big wood trusses.

DE VIDO: There I was looking into the regional vernacular, which tends to have exposed structure. We were trying to recall mining camps, the Old West, and all that (Fig. 2.8).

THORNTON: A bold, gutsy structural temple would not go over so well on Fifth Avenue.

DE VIDO: That's for sure. Another thing you have to think about

Figure 2.7 The compact steel details seen here engage steel tendon bottom chords of queen post trusses.
Robin Hood Dell, Philadelphia, Pa.; Architects: *MacFadyen/De Vido*; Engineer: *Lev Zetlin Associates Inc./Thornton-Tomasetti.* (Photo: *Lawrence S. Williams.*)

with exposed structure is insulation. A good option is putting it on the outside. This works very well since it lessens thermal movement in the roofing membrane, thereby lessening the possibility of ruptures and leaks.

THORNTON: A lot of the 1960s and 1970s university work was exposed concrete mixed with brick, where the brick sits on an exposed beam—the eyebrow where the spandrel sticks out. There's a problem because the slab extends out and you get a thermal bridge where a cold spot develops. You have to be concerned about whether the area near the outside is going to be cold.

DE VIDO: Another thing you have to think about is controlling all

Figure 2.8 Wood trusses with exposed, bolted connector plates recall the rough-and-tumble of Wyoming's early days, while carrying the heavy snow loads of this ski area. **Snow King Inn, Jackson Hole, Wyo.**; Architects: *MacFadyen/De Vido*; Engineer: *Lev Zetlin Associates Inc./Thornton-Tomasetti*. (Photo: *Ernest Silva.*)

the mechanical tack-ons that buildings require. In Allendale, New Jersey, we did a simple industrial building, a modular distribution center. The developer wanted a better looking product because he noticed that his competition was getting higher rent for attractive buildings with good landscaping and controlled graphics. Normally, various trades will place their devices without coordinating with one another. Alarm people will put their conduit and boxes in one place, the utility company will put their meter in another, and the electrician will run the conduit in still another place. It just happens. We knew the structure would be exposed for reasons of economy and mechanical items would not be covered up. We therefore coordinated with mechanical trades at early meetings for no extra money.

THORNTON: What you're really saying is that it takes more attention to all the other systems in an exposed structure, and they potentially could get more expensive as a result.

DE VIDO: Potentially. It doesn't have to be if you have the proper coordination between all the disciplines.

THORNTON: When all the piping and sprinklers and mechanical systems are exposed, then the architect gets involved in the design of all ductwork. The ductwork may go from a relatively inexpensive item to a very expensive item.

DE VIDO: It's important to ask the right questions early enough. You have to talk to your engineers, find out right away what is more economical and try to work within that framework rather than making arbitrary early commitments to, say, round ducts. At minimum, you should ask what is more economical, round or rectangular? And if the answer is rectangular, ask what you can do with the hangers and flanges to make them tidy. It's also important to locate them, because if you leave it to the mechanical contractor, the contractor may not be sensitive to the appearance.

THORNTON: This relates to the idea that when the structure is covered, the structural module doesn't have to bear any relationship to the ceiling module, the partition module, or whatever. But the minute you expose the structure, then it, too, has to fit into an already complex situation. Do you think that ends up costing you more time in designing?

DE VIDO: There's little doubt that it throws a lot more work on the designers, and, yes, they should get paid for it. The owners have to be convinced that you are going to save them money by doing more design work. An enlightened client will grasp the idea.

THORNTON: But when you come into a project and it is a Snow King or the Robin Hood Dell Pavilion in Philadelphia and you negotiate your fee—generally, there is little known at that time about what you are going to do.

DE VIDO: That's right.

THORNTON: So if you tell your client that you are fairly well known as an exposed structure advocate and it's going to cost more in the design stage, it's very difficult, even if in the long run it will save money.

DE VIDO: Yes, very difficult. The client will weigh your fee against your competition's. It's hard to deal with. But you can sell it on two grounds—on the basis of economy, which everyone will always listen to, and by pointing out that it can form the building's decorative, daring, exciting elements.

Another factor that can be important in certain types of structures is ease of inspection. In an exposed long-span structure, for example, you can look up and see if water is getting at it or if cables are beginning to slacken.

THORNTON: There is a hesitancy to use an exposed steel frame in parking structures because people think steel tends to corrode. But the beauty of exposed steel is that if it does corrode, you can instantly see it. In a posttensioned concrete structure, salt can get in and start eating at the tendons and you won't see it until the slab starts sagging and deflecting—and then you're in danger.

DE VIDO: Another issue is public acceptance. The public usually accepts exposed structures if they are attractive, but it depends who is looking at it.

THORNTON: Do people ever say that it looks unfinished? When are you going to put the rest of the building on?

DE VIDO: Some do, but it depends how the building is designed. People accept it in places such as airports, big public buildings, exposition halls, that sort of thing. But in office buildings or houses you may run into some resistance.

THORNTON: The criticism of a lot of the exposed concrete structures in the 1960s was that they were cold, impersonal. Marcel Breuer's work was called brutalism.

DE VIDO: LeCorbusier was more brutal than Breuer. Breuer did a lot of precast, as did Walter Gropius. I worked with both Gropius and Breuer. It is more daring and difficult to achieve a good building with exposed structure, but I find it very satisfying. It's a way to meet the public's desire for relief from the sameness of unrelieved sheet rock walls and monotonous strip windows. The postmodern movement added decoration, and its leaders did it in a sensible, intellectually satisfying way, although some did it badly, using awkward appliqués. I find it much more satisfying to expose the structure because not only is it a necessary and integral part of the building, but it can be more appealing in appearance. And there is ample historical precedent from classical Greece and Rome to Romanesque to Oriental styles—they are literally loaded with exposed structure, the very rich sort.

JAMES INGO FREED

To architect James Ingo Freed, structure defines a society. It should be an integral part of the creation of architecture, he believes, because that's what will be left after the passage of time. Freed, a partner in Pei Cobb Freed & Partners, New York City, created the vision for the expansive space truss of the 1.7 million-square-foot Jacob K. Javits Convention Center in New York City. It brings to life his concept of using structure to create spaces, not just surfaces, a concept he is also putting to work in the expansion of the Los Angeles Convention Center to 2.5 million square feet. Among Freed's most notable recent projects is the U.S. Holocaust Memorial Museum in Washington, D.C., where he says the structural language was used for things that go beyond structure, for the power to be evocative.

TOMASETTI: Jim, how do you define exposed structure?

FREED: Exposed structure is anything holding a building up that is used architecturally and artistically. Taking the structure and the needs of the building and turning them into a "type" of architecture...that's not just a structure; it is structure as an architectural essence. It elevates technology to art and can become mythic when you succeed.

Suppose you have a bond issue for a public building that's going to cost, say, $140 million. By the time the preliminaries are in place, the money value has gone down and the agenda has gone up. That seems to always happen. So you end up having to start design with about half of the real program. But you can see the handwriting on the wall—everything is going to disappear, everything except the structure. Structure can't disappear. Structure has to stay, or the building won't stand. So you ask yourself, how can I use structure to create architecture? You want to have a decent building that gives you a defined space and form as well as a sense of quality.

Actually, structure has the power to be evocative. Take, for example, the structure for the Holocaust Museum. It twists and turns. It's built in ways that are not necessarily the way you

---

would build it if everything did not show (Fig. 2.9 *a,b*). What this leads to is that structure can embody meaning. Structure gives a building a sense of unity, a sense of reality. In the Holocaust Museum we took off from structure and invented a language to develop everything. The techtonic language uses a vocabulary of doubled angles that forms doorways, windows on the concrete floors and on the ground floor—they tie everywhere. In this sense the structural language was used for things that went beyond structure.

TOMASETTI: You're talking about the Holocaust Museum. Have you applied this concept to your other designs?

FREED: Perhaps. The Jacob Javits Convention Center is a somewhat different case in point. It was perhaps the most interesting challenge up to that time. We actually had no time for design. The structure was not just a typical space frame structure. It was 5 feet by 5 feet by 5 feet, as submembers of a 10-foot by 10-foot grid. What we did was to make that whole structure work as one unit. Everything was shaped out of one thing. The periphery at the floor is a sort of concrete table, and the tiers higher up are carried by 90-foot spaced columns (Fig. 3.25). Ninety-foot spaced columns are not small; they are quite big. How do you take a space like that and give it a horizontal dimension? How do you make each of these spaces read? I decided to do so by skewering space vertically. That means that I used a vertical centralizing axis, and each of several "different" spaces, which were meant to be horizontally transparent, had vertical "skewers of space" everywhere. When you come into each space, you know it's part of the whole (because of the transparency), and yet it has its own identity.

TOMASETTI: I asked about your definition of exposed structure because often in our industry there are people who think that exposed structure means you have to touch the steel or the concrete. I think you're talking about a more general definition in which anything that expresses the form of the structure in a building is considered exposed structure. It can be covered with different materials, but it's still exposed structure.

FREED: Although I try very hard to expose as much as I can, in the end you have to cover something. But I think you're right. If you look at the museum, for example, the mullions are not the exposed structure, but they speak of a structure as if it were a surrogate. It's a virtual situation. We have virtual structure as well as real structure.

TOMASETTI: I gather from what you say that you prefer to show the

Figure 2.9 (*a*) and (*b*). Here the exposed structure is used as a way of evoking an era and a system of twisted brutality. Note double angles on walls in both construction photos.
U.S. Holocaust Memorial Museum, Washington, D.C.; Architect: *Pei Cobb Freed & Partners*; Engineer: *Weiskopf & Pickworth.* (Photos: *Alan Gilbert.*)

(a)

(b)

material of the structural element itself as much as possible.

FREED: I do prefer it. I am more comfortable with that approach because you don't have to begin by saying, how do I use something to stand for something else? There is the problem of the curtain wall: the curtain wall stands for something you don't see. I would like my next building to have a curtain wall that has all the structure on the outside. It should be inverted with the structure of the curtain wall facing outside and glass on the inside.

TOMASETTI: You can't tell us what this project is?

FREED: I cannot talk about it. However, there is a project in Washington, D.C., that I would like you to see. It's a private building called 1299 Pennsylvania Avenue. It has a structure that is almost unbelievable. You know, in Washington you cannot build above 110 feet. The typical block is so large you absolutely have to have an open space in the middle of it. The open space in this case was extraordinary because it moves in many directions at once. But at the same time it's quite logical. Its structure is in a sense exposed. The base of the building within the building is about 95 by 60 feet, and the structure is highly stressed. There is a 15-foot space in front of it that acts as a tension ring, and the ring is what holds the building together. It is then glazed as simply as possible. The result is that you see the structure through the glass; it has to be seen that way because it's too large to be seen any other way. Imagine a building being held up by columns that tilt forward and backward, and the only way to keep it from flying apart is the 15-foot-wide tensioning ring around it (Fig. 2.10).

I am interested in structure, but not for structure's sake. Structure helps me do what I like to do best, which is to make spaces. I think of space as the interval between structure. You have space swirling around, and then you come to a wall and you no longer have space; you just have a surface. I feel very comfortable doing things in the void, in working in the in-between space. But you need a measuring device, and structure is that device. It's the thing that says, oh yes, it's about so high up to the roof and about so wide.

TOMASETTI: That's interesting because it reminds me somewhat of the Oriental philosophies in which the emphasis is on the nothingness of some things, the space between things.

FREED: I am interested in the void and voids that are perforated. There are some spaces you can walk into where you actually feel the palpable presence of spatial movement.

Figure 2.10 This midblock atrium uses exposed sloping structured columns to create excitement.
1299 Pennsylvania Avenue, Washington, D.C.; Architect: *Pei Cobb Freed & Partners*; Engineer: *Tadjer-Cohen Edelson Associates, Inc.* (Photo: *Joe Aker.*)

TUCHMAN: Like the space behind you. Look at the photograph on the wall.

FREED: That's the Holocaust Museum. It's a structure developed from a hexagonal whole (Fig. 2.11). The hexagon is relatively stripped, but I wanted to enter the corners. As this corner cut back, I was surprised to find sloping surfaces that became triangular voids. I did not make those triangles (they are really not triangles), but that's what happens when you take independent structures at the hexagonal point and tilt them inward until they meet. This is an all concrete beam structure that carries a wall. Here is the dome, and in the middle there is a clear arc that lets the sun draw an outline across it. And here is where people just sit—it is a space for contemplation.

The inside of the building is perhaps more interesting than the outside. As you go in, one finds a shell, with stairs that service this whole space. These stairs are an interesting exercise. Look what happens—the geometry is so calculated that it is its own powerful thing, but you don't see it. In this case, I believe the skin had to be pure because it had to function as a structural zipper.

TOMASETTI: Jim, when you start a project, at what point do you interface with the structural engineer?

FREED: Very early. For example, on the Javits Center I got involved with the engineer right at the beginning. Matt Levy was the engineer, and I explained to him that we were going to do a space frame that was not like a neutral space frame. I told him that it had to be 45 degrees. He said that all space frames he had worked on were 60 degrees. I told him no, that this had to be 45 degrees because I wanted to carry it all the way down to the walls (as structure), and the only way I could do that was to translate the corners at 45 degrees. With any other angle I would be in trouble.

TOMASETTI: You have a big problem going from horizontal to vertical if you use 60 degrees.

FREED: Yes. I had the talk with Matt right at the beginning to make sure we could do it; Matt then came up with another idea. He said that we couldn't do a pure space frame, but we could do a space frame with a truss system.

TOMASETTI: I know you have done work in California, such as the Los Angeles Convention Center (Fig. 2.12). What is your experience there? The engineering community tends to be a little more demanding when setting structural requirements because of understandable concern about the seismicity problem.

FREED: What you say is interesting because what we found is that you have to design even higher, to a seismic (Richter magnitude) 7, not a 5. We used base isolation in a San Francisco library building that is about seven to eight stories high and about 350 feet long by 220 feet wide (Fig. 2.13). We put plates in the sidewalk with other plates that would kick up and open up when the base needed to be isolated from the building. Since the actual foundation had to be able to move in every direction, the only

Figure 2.11 This construction photo shows exposed concrete lintels carrying an upper stone ring pierced by triangles.
U.S. Holocaust Memorial Museum, Washington, D.C.; Architect: *Pei Cobb Freed & Partners*; Engineer: *Weiskopf & Pickworth*. (Photo: *Alan Gilbert*.)

Figure 2.12 Long span spaces and large volumes make prime candidates for exposed structure.
Los Angeles Convention Center, Los Angeles, Calif.; Architect: *Pei Cobb Freed & Partners*; Engineer: *John A. Martin & Associates*. (Photo: © *Thorney Lieberman*.)

way we could accomplish that was if the little doors popped open.

TOMASETTI: You expressed the base isolation so that when people come to the building they get the feeling of floating?

FREED: We had no choice. You couldn't get in and out of the building otherwise.

TOMASETTI: Maybe necessity is the mother of invention. I don't think anyone has thought of expressing the base isolation on a building in that manner.

TUCHMAN: We've been talking a lot about spaces and structure. I would also like to get your thoughts about exposed architectural concrete.

FREE: I love architectural concrete. I like it because it does everything at once. It gives you finish; it gives you shape; it gives you structure. Nobody can do it right today because it seems we can't afford it. You can't get architectural concrete in the States any more. You can get pretty good precast. That's what I used on 1299 Pennsylvania Avenue, and it looks like limestone now. However, I prefer architectural concrete because what you see is what you get. The problem now is that contractors move very fast to make a profit, and you cannot move so fast with architectural concrete. There are so many little things that can go wrong and cause imperfections. You need a lot of patience with architectural concrete. You really need to look at it, analyze it, and take the necessary time to solve issues, and the contractor has to be willing to do the same.

TOMASETTI: Do you think the fact that it's harder and harder to work with exposed concrete is going to affect the type of architecture we're going to see in the future?

FREED: You can probably still ask for a concrete building and get a pretty good one. It still pays for people to spend some time to do it right. But you really need more than that, in fact you need commitment to do it really well.

TOMASETTI: The downturn in the business cycle might work in its favor.

FREED: That's right. When everybody is busy, they don't spend any time doing things. I used to see a plasterer take his family out to look at a flat plaster wall because it was so beautiful. Now it's just "slap it on" and get done with it. The process has become more important than the product. That is where I think structure and architecture will face a problem because exposed steel columns and exposed concrete columns take more time and attention. Nevertheless, structure is always going to be the essence of a building. After you take everything away, you still have structure.

Figure 2.13 Base isolation protects this building from earthquakes. The need for a gap around the building will be clearly expressed.
San Francisco Main Library Expansion, San Francisco, Calif.; Architect: *Pei Cobb Freed & Partners*; Engineer: *OLMM Structural Design.* (Photo: *Gerald Ratto.*)

TOMASETTI: As a structural engineer I am glad to hear you say that.

FREED: But it's true, isn't it?

TOMASETTI: Of course.

FREED: If you say that structure is primary to a building, you could say that the mechanical aspects are secondary. The old classical example was to build walls that were 40 feet apart because that was the dimension needed to bring in light and air movement. That made for a certain structure, but it didn't make anything that was very exciting. My job is to do things with that structure that you could never do before. Today, spaces are just incredible; spans can be unimaginably long. I don't think big is necessarily good, and I don't think that truss work is necessarily good. What I do think is necessarily good is the ability to think in terms of larger scale when you need to do it, to be in the scale of the city.

TOMASETTI: Do you get involved in using exposed structure as an accent, such as a dome? Do you like to do that?

FREED: I like to use structure in the way that it's best used. For example, the Holocaust Museum has brick walls on the surface. Brick walls today usually have a structure behind it; you probably have one-wythe-thick brick walls and that's it. When you see it, it looks artificial. When discussing the museum, I knew we couldn't have bricks like that, but we could have the brick bear itself—not bear the building, but bear itself. If the brick does that, then you can use the bricks in a structural fashion. When bricks begin to be load-bearing, they begin to be a structural element. Brick that is hung and glued never looks alive, but when you compress it, it does.

So the museum has 8- to 12-inch-thick walls that carry their own weight. We used a computer to lay out a pattern of English bond, which happened, I found out later, to be the actual pattern used at Auschwitz. Every brick was called for and put in position; not one brick was left to chance. I also used the computer to add the stiffeners and the angles to be sure they worked perfectly. Every place there was an intersection, there was an angle. That's the way in which the brick was used structurally. I would be lying if I said I didn't want to use the brick for its textural possibilities and form as well; the form of the building all flows from that. But at the same time I always feel that somehow you gain something extra if you don't deny structure. Denial of structure is what I think we have today. Many artists deny structure; they divorce themselves from it and think that it ought to be hidden. They feel ashamed of it; they want walls. I can understand that, but I don't see why you have to get rid of one for the other. In fact, I think that surfaces that do not grow out of structure (in its largest sense) lack the spirit of authenticity.

TOMASETTI: I want to ask you whether you think in the next five or ten years we're going to see more or less of architects' exposing structure the way you have always done it?

FREED: I think the most interesting work done now uses exposed structure in a much more complicated way than I do. I think it's used to mean something as well as to be something. I think that structure will be more to look at rather than just be a measure of space. In the end structure has to be the only thing that works because it's the only thing from which you can't take anything away.

TOMASETTI: Great examples of that can be found walking around Rome. They haven't refinished buildings; they haven't added finish to the original structure. They're just natural structures that have been there for 600 or 700 years.

TUCHMAN: What's left after 600 or 700 years?

FREED: The structure. That's why a structure defines a society, I think.

TUCHMAN: You mentioned something before about the difference between the engineering aesthetic and the architectural aesthetic. What did you mean by that?

FREED: To my mind the engineer tries to design the most economical structure possible; he's charged with that. However, to me as an architect, the most economical column may make me feel uncomfortable and perhaps it also does not define the space properly. So I come along and tell him to make the column a little bigger. What I've found is that any real structure seems always less heavy than I think is necessary. Lots of architects take another view. Some engineers even want to make it very economical and aesthetic, too. But in today's world, most engineers are concerned about the least amount of steel they can use for this and the least amount of concrete they can use for that. So the architect says that we can use it here, but we can also make it feel aesthetically comfortable as well as speak to the eye in a convincing way.

GYO OBATA

Where else would the authors interview Gyo Obata but in Union Station? Hellmuth, Obata & Kassabaum restored the turn-of-the-century St. Louis landmark, where the graceful exposed lattice truss roof that once covered train platforms now protects a popular shopping mall. Obata, chairman and design guru of the large St. Loui-based architect-engineer, initially didn't think of himself as practitioner of exposed structure. But upon reflection, he found many examples of it in his work—each with a clear motivating factor.

OBATA: There are a lot of architects who feel that structure shouldn't be expressed. A building should have a finished covering on the outside and a finished covering on the inside. But I believe that if it's appropriate, aesthetic, and solves a problem, then the structure should be expressed. But you need to have a reason to do so—a very tall building or a very long span.

TUCHMAN: Have you ever done an exposed office building?

OBATA: That's more unusual. About 15 years ago, on the 50-story First International office building in Dallas we did express the wind bracing. When you're that tall, it really helps because it's the wind that you're concerned about. To provide the necessary support, cross bracing was used along the exterior walls of the building. A benefit of this kind of application, by the way, is the reduction of the number of pounds of steel required, which then reduces the building's cost.

In this case, the structure was expressed on the inside—some of the rooms had diagonal bracing going through them (Fig.

2.14). I also lit the whole thing, so that at night the structure could be seen (Fig. 2.15). But during the day it was hidden. I covered the bracing with a glass skin. On a 20-story building it would be very difficult to express the structure architecturally because the frame is pretty simple. But once you get to 60, 70, or 80 stories, you can begin to work with some expression, as was done in the Hancock building in Chicago (see Fig. 3.13).

TUCHMAN: The British architects are using exposed structure quite extensively.

OBATA: Yes, Norman Foster and Richard Rogers have done exquisite factories that are hung on cables, with all the structure exposed. They evidently had clients who wanted their factories to make a kind of statement. Another good example is the Georges Pompidou Centre in Paris that has all the structural and mechanical systems on the outside of the building (see Fig. 2.30). It's not really high-tech in a sense, but it's designed with a high-tech imagery, and a lot of it is very aesthetic. Instead of doing a lot of stone detailing, as on some of our postmodern buildings (see Fig. 1.9), they're using high-tech imagery but doing equally complicated detailing. They make it look as if it's all prefab, but it isn't. It's really very complicated and expensive, too.

TUCHMAN: Can exposed structure ever be a vehicle for reducing costs?

OBATA: We did a very inexpensive building for the Missouri Botanical Garden in St. Louis that has all exposed structure on the inside. Called the Ridgway Center, it's the main public entrance to the garden. A barrel vaulted skylight delineates the entrance and serves as the focal point for the rest of the building. The center features a bent steel frame where all the steel is exposed and has the skylight over it. It was really high tech in a sense. We did custom detailing to bring the structural steel within the budget.

TUCHMAN: How much synergy is there between architect and engineer in an exposed structure?

OBATA: It's very important to work closely with a creative structural engineer in exposing a structure. I worked very closely with Bill LeMessurier on the Air and Space Museum, for example. We had great walls of glass in three galleries that faced out into the mall, and it was all skylit above. To provide the necessary supporting structure for the glass walls and skylights, Bill came up with a triangular truss system consisting of five tubular steel trusses (Fig. 2.16). As well as supporting the glass, these L-shaped bent trusses also support suspended objects weighing

Figure 2.14 Wind bracing is exposed to the interior.
First International/Interfirst II, Dallas, Tex.; Architect: *Hellmuth, Obata & Kassabaum, Inc.*; Engineer: *Ellisor & Tanner, Inc.* (Photo: *Barbara Martin.*)

Figure 2.15 Wind bracing behind an all-glass facade is expressed by lighting. **First International/Interfirst II, Dallas, Tex.**; Architect: *Hellmuth, Obata & Kassabaum, Inc.*; Engineer: *Ellisor & Tanner, Inc.* (Photo: *Barbara Martin.*)

up to 4 tons (Fig. 2.17). All this structure in the room was expressed from the inside in the open galleries.

Exposing a steel structure on the inside is one thing, but exposing it on the outside is something else. If you cover a structure and control the temperature, you avoid dealing with what can be a large amount of movement. In a climate like St. Louis's, where you have temperatures from below zero to over 100°F, that movement can cause tremendous problems.

TUCHMAN: Have you ever exposed a steel structure to the elments?

OBATA: When Corten was very popular, we used it in the exterior columns of the Emerson Electric corporate headquarters. When the structure went into the interior of the building, then we fireproofed it. Because Corten rusts, dark brown paving or other building materials must be used to conceal the resulting dark brown stains. In the case of this building, we used dark brown brick pavement.

TUCHMAN: Might you be more likely to express the structural system without exposing it to the elements?

OBATA: I have done that. At Lambert–St. Louis International Airport, which I designed with Minoru Yamasaki, we expressed the arches on the exterior, but protected them by covering them with a copper roof. We also had to take extra care with how the span came down, how it met the ground—because it's all visible.

TUCHMAN: Have you exposed concrete?

OBATA: I have done thin-shell concrete projects such as the St. Louis Science Center, the former McDonnell Planetarium. The concrete is the finish material although it is covered by a base coat of neoprene to waterproof it and then by Hypalon. Hypalon, which was then only available in a liquid form, protects the neoprene from ultraviolet rays and provides the color—in this case a shade of white. The same thing was done at the Priory Chapel southwest of St. Louis. This Benedictine chapel features a circular design and two tiers of 22-foot and 18-foot-high parabolic arches topped with a narrow bell tower of 10 more arches (Fig. 2.18). It made the cover of *Life* magazine right after it was finished in 1962. It has a very thin shell, only 3 inches of concrete. In these examples, I exposed the concrete for its sculptural qualities (Fig. 2.19).

The new church we are designing for the Latter Day Saints in Independence, Missouri, will be shaped like a seashell. I was thinking of it in terms of exposed concrete, but it looks more and more as if it will be a steel structure with stone cladding and a metal roof.

TUCHMAN: Does an exposed structure have an element of the daring?

Figure 2.16 Primary triangular trusses are vertical at the exterior wall.
National Air and Space Museum, Washington, D.C.; Architect: *Hellmuth, Obata & Kassabaum, Inc.*; Engineer: *LeMessurier Consultants.* (Photo: *George Silk.*)

Figure 2.17 Primary triangular trusses turn at the roof to hang airplanes.
National Air and Space Museum, Washington, D.C.; Architect: *Hellmuth, Obata & Kassabaum, Inc.*; Engineer: *LeMessurier Consultants.* (Photo: *George Silk.*)

Figure 2.18 Parabolic arches 3 inches thick form a vaulted chapel and bell tower. **Priory Chapel, near St. Louis, Mo.;** Architect: *Hellmuth, Obata & Kassabaum, Inc.;* Engineer: *Paul Weidlinger.* (Photo: *George Silk.*)

OBATA: I don't believe that structure should be exposed just to be daring, although it can be exciting and interesting. I think that with every structure, the architect has to use judgment about what is really appropriate for that particular project.

The spiral shell design for the church, for example, really grew out of that congregation's desire for an architectural form that was different from any other church. I felt that because the shell was a universal, natural form that's found all over the world, it would be in keeping with their plans to send their missions all over the world. I thought it made a good symbol, and they agreed.

Figure 2.19 Arches form the interior architectural expression of this vaulted space. **Priory Chapel, near St. Louis, Mo.**; Architect: *Hellmuth, Obata & Kassabaum, Inc.*; Engineer: *Paul Weidlinger.* (Photo: *George Silk.*)

CESAR PELLI

The conference room of Cesar Pelli's office in New Haven, Connecticut, overlooks the exposed load-bearing stone structures of Yale University—quite fitting because Pelli was dean of the College of Architecture there from 1976 to 1984. Over lunch with the authors, Pelli explained that after starting his career in the office of Eero Saarinen, he moved to the West Coast. He was director of design with Daniel, Mann, Johnson & Mendenhall and then a partner for design with Gruen Associates, both in Los Angeles. Pelli says he came to New Haven prepared for a slower kind of life and practice, but that was not to be. He hadn't been dean a month when he was selected to do the expansion and renovation of the Museum of Modern Art in New York City. That launched a highly successful practice that has designed the World Financial Center in New York City and a proposed 1999-foot-tall tower in Chicago.

PELLI: I have thought a lot about the problem of exposed structure because it's an issue that's of interest to architects always. As happens with many things, you tend to reach a working conclusion that allows you to keep going, and then you don't think much more about the problem. But it's a very good problem.

You can talk of exposed structure from different levels. At one level are utilitarian structures that are the simplest form of covering—a shed for a horse, a shelter above an excavation. Logically and economically, those structures are exposed. But it is quite another thing when you start to express the structure in a large complex building. The problem has to do with the nature of our buildings. Today we build self-contained, airtight, controlled environments. We want to keep not only the rain, the dust, and the thieves out, but we also want to control our humidity and temperature precisely, no matter what is happening outside.

To accomplish that, we have developed very good skins to wrap our buildings in. Even the fireproofing on steel is a kind of skin. We need to cover our buildings with a sort of raincoat. One can keep the structure or pieces of the structure outside the raincoat, but that's doing it the hard way. You will have one piece of the structure fluctuating in temperature and the others not. Also, with one piece of the structure exposed to rain, wind, and water, at some point you need to thoroughly isolate it, which, if it is steel, is very difficult to do.

At the John Deere headquarters in Moline, Illinois, Saarinen did it well. He managed to get a reprieve from fireproofing by having the structure at a certain distance from the building. But the structure inside the building still has to be fireproofed. So at some point of going in and out, you need to seal it, which is not a natural thing to do.

So by and large, we end up covering everything—both structure and spaces. In winter, at the exposed concrete British Arts Center in New Haven, I have touched the inside of an outer column, and it is cold—you can feel it—and a bit humid. Concrete is dense, and the transmission of cold and humidity may be slow, but they are transmitted.

THORNTON: And it stays cold for a long time.

PELLI: Exactly. Now, if you cover your structure with a raincoat, the question becomes, do you design that exterior raincoat with changes of materials so that the structure inside is visible or expressed in some manner, or not? As you can see, the problem is quite different in different building types. If you are doing hangars or large open spaces like the Winter Garden at the World Financial Center, it's rather easy because you are using special forms (Fig. 2.20).

THORNTON: It becomes almost natural to expose it.

PELLI: It becomes difficult not to expose it. But the crux of the issue is the tall office building. There the expression of the structure exists to me at two levels. One is to design it so that the form of the building and the form of the exterior are one. Or will you express only the principle of the structure, not its form?

In a way, that's what Mies van der Rohe did at the Seagrams Building. He put those mullions on the exterior that look like structural elements, but he did not put one per column. He put one every 5 feet. The column is not exposed, but as you see Seagrams, you have no doubt that it's steel construction he is expressing. He is expressing the principle, but not the particulars of the structural form.

However you cover it, however strong or weak the signals you give about the structure inside, what is most important to me is not to give the wrong signals. That's why I find myself so uncomfortable when I see the columns of the AT&T Building in New York City. They are done as if they were stone bearing columns, but they look like toothpicks supporting a 36-story building (Fig. 1.9).

THORNTON: You know they are not stone columns.

PELLI: It's trying to tell you, "I am a stone column." So for a millionth of a second, you accept that pretense, like in a movie. The actor tells you, "I am blind, or I am the King of Norway." And

Figure 2.20 With a glass skin, the Winter Garden has presence as an overall shape while still exhibiting exposed structure.
World Financial Center, New York, N.Y.; Architect: *Cesar Pelli & Associates*; Engineer: *Thornton-Tomasetti Engineers*. (Photo: *Timothy Hursley*.)

you believe it. And when a part of you believes that it is a stone column, suddenly you feel that the whole building is going to crumble because it won't take the load, and you feel very uncomfortable. The most critical thing to me is not to misrepresent.

Through the centuries there have been many wonderful buildings that made no effort to express the way they were built. But also until the 19th century, there were few efforts to disguise the basic system. Architects have always disguised specifics. They have hidden beams, put in false columns.

THORNTON: If you think about it, until concrete and steel arrived, it was mostly masonry. You couldn't disguise it totally.

PELLI: But you could always add a column that wasn't due on the job or add a piece of plaster that looked as if it were stone.

THORNTON: Like ornaments?

PELLI: No, not ornaments. Using forms that looked structural but were not structural, or making pieces of the structure disappear and others not, so you would believe there was no structure there.

THORNTON: In the early 1900s, designers didn't want to let go of heavy walls. So they built with them although the building was actually supported by a steel frame. They said the heavy walls weren't doing anything, but they actually could have.

PELLI: Exactly. Eero Saarinen had a nice anecdote about that. While traveling in Italy he had stopped in a small town to chat with a stone mason. Eero explained that he was an architect and what an architect does. At the time, he was very proud that he was designing the CBS building in New York. He said, "I am doing a 36-story building." The man looked at him, thought for a second, and said, "Thirty-six stories? You mean the ground floor has to be solid stone?" He may not have had a college education, but he had a clear understanding of the limitations of stone construction.

But to go on with my thoughts about classification. As I was saying, you can express a principle, like Mies, or you can also express a structure specifically—more like the John Hancock tower in Chicago (Fig. 3.13). I remember that at the time before John Hancock was built, a friend at Skidmore Owings & Merrill used to say that they were doing Mies one better.

THORNTON: More Mies than Mies.

PELLI: Exactly. They were really expressing the structure. But to me it was a misunderstanding of Mies because Mies never was interested in expressing the specifics of the structure—only the idealized conception of it. I think they missed the point.

They are two very different things. To express a structure would be a reasonable solution. Another one is to design unusually complicated structures so you can feature the structure as a prime architectural element. There is still some of that going on in Europe. But here, the basic American pragmatism and indeed developers' and owners' concern with economy and efficiency have made those explorations very rare. Even then, it has all been done within rather tight economic parameters, right?

THORNTON: That's right. There is the work of Nervi and Felix Candela. But the reason it never caught on in the United States is that nobody was willing to pay for it. Only at a World's Fair or special event can you find a benefactor to support that type of structure. On the other hand, at a large commercial structure like the World Financial Center, the need to have a central meeting place stimulated the Winter Garden. In a way, the Winter Garden does have a raincoat. The skin covers it outside, nothing sticks through, and yet it really is still an exposed structure (Fig. 2.21).

PELLI: Absolutely. But it remains quite rational. One could have conceived of a structural system that was all outside with the glass skin on the inside. That's a form-giving element at a cost. But even in the Winter Garden as you know, Olympia & York was willing to pay for this special space, but every piece of it kept on being investigated to make sure monies were not being spent unnecessarily.

THORNTON: On the Miglin-Beitler building in Chicago, one of the things that makes the solution quite obvious for the engineer is that you and your team recognized that the building was special and planned around the fin columns. As you recall, we were able to widen the structural footprint of this tall, narrow building by projecting fin columns beyond the lot line and having them bare at the sidewalks. You saw the economic benefit of this structural approach and adapted the fin columns to create the expression (Fig. 2.22).

PELLI: Exactly. I'm of a very pragmatic bent because I like for my buildings to get built. I recognize that in a building that tall, the structure has to take precedence. As soon as you decide to do it, even before you know what it's going to look like, you have to know that the approach is intelligent, that the building has a rationale for being what it is—not arbitrary or capricious—and that it is economical. A building that tall is only going to get built if we are clever enough to reduce or eliminate all unnecessary costs. As long as it is handsomely proportioned with a beautiful silhouette, it doesn't need secondary gestures. The beauty is in the silhouette and the height. Both for rational purposes and economic purposes, we had to start from an intelligent and reasonable structural system.

THORNTON: There are architects that make engineers go through gymnastics to fit the structure into some preconceived architec-

Figure 2.21 The Winter Garden interior plays trussed barrel vaults against the soft lines of palm trees.
World Financial Center, New York, N.Y.; Architect: *Cesar Pelli & Associates*; Engineer: *Thornton-Tomasetti Engineers.* (Photo: *Timothy Hursley.*)

Figure 2.22 Fin columns emphasize the verticality of this tower while providing the advantage of a wider structural base.
Miglin-Beitler Tower, Chicago, Ill.; Architect: *Cesar Pelli & Associates*; Engineer: *Thornton-Tomasetti Engineers.* (Photo: *Kenneth Champlin.*)

Figure 2.23 This proposed terminal has sweeping arches supported by intermediate steel towers.
Kansai Airport Terminal Proposal, Japan; Architect: *Cesar Pelli & Associates*; Engineer: *Thornton-Tomasetti Engineers*. (*Photo: Cesar Pelli & Associates.*)

tural concept. There are buildings where no columns that start at the top make it to the bottom without being shifted. In the end, the buildings don't turn out to be as good as they could have been.

PELLI: I agree completely.

THORNTON: Your competition entry for the Kansai Airport design was a bit of a diversion from your normal style. And it was truly an exposed structure. Does that indicate that you might like to do more of that?

PELLI: I have always been interested in working on structural systems that one may expose or take advantage of. But I believe one should use them in a rational manner. Kansai was a great opportunity. I liked it because from the beginning we opted for a structure that created some ambiguities that required therefore a very clear intellectual handling. We chose to do single arched forms, but with multiple supports. It was not an arch, but rather an umbrella or parasol form (Fig. 2.23). There are innumerable bridges and causeways that have an arch form because they need to clear a certain height. And in classical architecture there are arches with intermediate support that are not really arches. Those are good precedents. The interesting thing to me was how to design it so it is indeed light, feels light, remains light. We only loaded it with precast pieces and it is expressed both inside and outside.

KEVIN ROCHE

The setting of Kevin Roche's office in a stately mansion on Lake Whitney in Hamden, Connecticut, offers visitors a hint about his views on structural expressionism. Roche has been involved in the design of some very well-known buildings that expose structure both currently and when he worked with Eero Saarinen in the 1960s. But he draws a sharp contrast between his passion for pure, simple, direct, elegant, and beautiful structures and his strong feelings against structural expressionism as a style or form of design. Roche explains that he does not believe the abstract concept of structural expressionism is an essential element in the design of a good building. There are so many other factors about why a building exists, he points out, and so many other considerations an architect must weigh—use, relationship, appropriateness, and urban design, among many others. He has used direct structure to make architecture in the New Haven Coliseum (Fig. 2.24), the Cummins Darlington Factory (Fig. 2.25), the Creative Arts Center at Wesleyan (Fig. 2.26), the Metropolitan Museum of Art American Wing (Fig. 2.27), and other buildings.

THORNTON: Do you think you could duplicate the TWA terminal today with the present economic climate?

ROCHE: The client support system which is necessary to build any building wouldn't indulge in such a building today. The design of the building is without concern for structural expression. It's just a wonderful piece of sculpture. It is a large sculpture that is supported, not a structure that gains its beauty from the elegance of its engineering. It is a form first and a structure second (see Fig. 4.1).

THORNTON: The Dulles terminal is the same in a sense, although a little more powerful.

ROCHE: Dulles is more securely based in structure. There are two points supported by columns sloping out with a catenary between them. But it is more than just a structure. Eero used the structural expression to make a powerful form statement—one that has to do with Washington as a capital and the terminal as an introduction to flight (see Fig. 1.7).

THORNTON: A European group, including Richard Rogers and Peter Rice, have been experimenting with exposed structures. They say they can do this in Europe because Europe is not as industrialized as the United States, because the bottom line is not yet as important there as here.

Figure 2.24 Steel trusses and spandrels and concrete columns are exposed on the building at right, while brick clads but still expresses corner shaft/columns at left. **New Haven Coliseum and Knights of Columbus, New Haven, Conn.**; Architect: *Kevin Roche John Dinkeloo and Associates*; Engineers: *William LeMessurier and Associates, and Pfisterer and Tor, respectively.* (Photo: *Kevin Roche John Dinkeloo and Associates.*)

ROCHE: They are right. Norman Foster couldn't have built the Hongkong and Shanghai Bank (Fig. 1.3) or Richard Rogers couldn't have built Lloyd's of London or Centre Pompidou (see Figs. 2.31, 2.30, and 3.29) in the United States, because the system here doesn't support that kind of work. Undoubtedly, the time will come in Britain and continental Europe when the bottom line will rule and buildings like that will then be impossible.

THORNTON: All our construction managers are teaching them the American way right now.

ROCHE: There's also an enormous difference between private and public owners. In France major projects are government financed: the Opera, the Louvre, the Grande Arche. It would be impossible to do that with private financing. And it isn't possible in our society where government spending on buildings is so tightly controlled—except when it's something for

Figure 2.25 Steel purlins, girders, and columns are exposed to define the facade here. **Cummins Engine Co. Plant, Darlington, England;** Architect: *Kevin Roche John Dinkeloo and Associates*; Engineer: *Henry A. Pfisterer and Associates.* (Photo: *Balthazar Korab.*)

Figure 2.26 Exposed structural concrete bands at roof and grade contrast with finished stone panels here. **Wesleyan University Creative Arts Center, Middletown, Conn.;** Architect: *Kevin Roche John Dinkeloo and Associates*; Engineer: *Henry A. Pfisterer.* (Photo: *Kevin Roche John Dinkeloo and Associates.*)

Figure 2.27 Exposed steel girders and concrete columns and balconies serve as a backdrop for artwork.
Metropolitan Museum of Art American Wing, New York, N.Y.; Architect: *Kevin Roche John Dinkeloo and Associates*; Engineer: *Severud, Perrone, Sturm, Bandel*. (Photo: *Kevin Roche John Dinkeloo and Associates*.)

Congress. Then they build in the most bungling and expensive way possible. Even institutional buildings, aside from a few like the Getty Museum, are tightly budgeted. Our work at the Metropolitan in New York is very budget-controlled. High-rise building, of course, is investment building...

THORNTON: Profit machines...

ROCHE: And that's a big difference from the Hongkong and Shanghai Bank, which is an expression of vanity. Vanity has produced most of the "great" architecture.

THORNTON: What about religious expression?

ROCHE: That's also vanity. Personal vanity has been the driving force for many of the exceptional buildings. While an element of vanity remains in profit machine building, the challenge of the architect is taking the limits—the site, the zoning code, the engineering possibilities, the owner-investor's requirements, the market requirements—and making a building that attracts the highest paying tenants. The architect knows that when you cover the structure, you avoid a lot of expense and a lot of technical problems.

THORNTON: Another project you worked on was the St. Louis Arch, which is not a building but is exposed structure. It's very impressive.

ROCHE: Now that truly is a structural expression, but again Eero uses it for a very powerful form. The engineering is very close to the form. The idea is a catenary inverted. It is a composite structure with a skin of steel plate and a concrete core. The laminated steel plate, which is part of the structure in the lower third, is filled with concrete; the upper two-thirds are just the steel shell itself.

THORNTON: Is the exposed material stainless?

ROCHE: Yes, stainless clad steel plate.

THORNTON: It's a spectacular and beautiful form.

ROCHE: I think it's a perfect combination of architecture and engineering. The two come together to create a beautiful moment, a very beautiful object. The Golden Gate Bridge is a marvelous achievement, but it has been rendered less by the addition of decoration. In the St. Louis Arch there is no decoration. Most of the great engineering works are recognized for their beauty because they are just that, just superb engineering. Of course, engineers aren't opposed to manipulating a design a little bit to make something more understandable.

THORNTON: So much for pure engineering.

ROCHE: I suppose in a high-rise building limited by the most stringent cost controls the purest engineering might result as everything has to be cut back to the absolute minimum. Depending on the courage exercised in refining and cutting back, the engineer might achieve the purest structure. But no one else will ever see the purity.

THORNTON: They probably come closest to the purest exposed structure in Chicago. Why?

ROCHE: Part of the reason is tradition. There still is fairly strong respect in Chicago for structure as the most important element in buildings (Fig. 2.28). There's the kind of discipline among engineers and architects which isn't evident in New York.

Chicago has had a very simple, straightforward approach to buildings, which partly was Mies's influence and partly the influence of the very rational kind of building in the late 19th and early 20th centuries. All of those great buildings, while they may be clad with decorated stone, have a very sound, sensible engineering premise behind them.

THORNTON: Is the real structure of a building often expressed?

ROCHE: First, expressed and exposed are two entirely different things. The real structure is not necessarily what gets expressed. Architects tend to be quite naive about what structure really is; for them it's a simple column, beam, cross bracing, arch, or dome, but the structure is infinitely more complex. Sometimes the full structure reasoning is beyond expression.

Structural expression as a formalized style is remote from true structure. If, for instance, we take masonry bearing structures, the building *is* the structure. The stress diagrams are not necessarily articulated in a masonry structure. They probably change according to the temperature. As we know, stresses move around in buildings.

In recent years there have been many efforts to analyze the stresses in Gothic architecture, but one suspects those stress diagrams are not the whole story. The builders undoubtedly knew what they were doing, because the buildings are still standing. But who knows how the structures are now, several hundred years later, responding to gravity, wind, and temperature?

THORNTON: Can't you express structural elements?

ROCHE: Even when you get down to the elements of the building, at best you are giving a very broad brush stroke: this is a beam; that is a column (Fig. 2.29). There are those who would attempt to show what happens in a beam and a column by fabricating things that get thicker and thinner and wider. But it is still generalizing. For me, expressing structure would be to take a minimal necessary structure and show it as just that. Who does that?

You can't express a steel building. You have to fireproof it. A steel building is behind a muffle of insulation. The best you can do is put another piece of metal or something outside that says: behind me is a steel column; or, behind me is a steel beam. It is a metal coating on top of some insulation, which is on top of the steel. In a way, the structure is expressed, but you are not *seeing* the structure. You are seeing the expression.

THORNTON: Do you believe expressing structure is a worthwhile aim?

ROCHE: There is no virtue in my mind in expressing the structure. It doesn't make the building better or worse. The structure's

Figure 2.28 The "supercolumn frame" which braces this tower against wind sway is clearly expressed on two elevations.
One North Wacker, Chicago, Ill.; Architect: *Kevin Roche John Dinkeloo and Associates*; Engineer: *CBM Engineers*. (Photo: *Kevin Roche John Dinkeloo and Associates*.)

Figure 2.29 This development expresses structure with framed-tube elevations and massive legs.
Pontiac Marina Project, Republic of Singapore; Architect: *Kevin Roche John Dinkeloo and Associates*; Engineer: *Weiskopf & Pickworth*. (Photo: *Kevin Roche*)

there to hold the building up and whether you see it or not depends on a number of other issues. The skin is there to keep the weather out. If you're rational, there are no frills except the paint on the outside.

THORNTON: As with the human body and our skin.

ROCHE: You don't see the bones unless you x-ray the body.

THORNTON: How about long-span structures?

ROCHE: There's a possibility of seeing the structure in a large enclosure, a long-span enclosure. And bridges, of course, are the best opportunity for expressing structure. Engineers don't respond, in either case, as imaginatively as they should. Many of our great bridges are encumbered by design elements that have nothing to do with fine engineering. I have always been impressed by highway engineers where the cost-effective mentality is at work. However, except in a few states such as California, it produces things of not great beauty, but nevertheless completely rational objects. The parts that have to be steel are steel. The parts that have to be concrete are concrete, put together in the most straightforward way.

RICHARD ROGERS

The authors caught up with Richard Rogers as he was about to speak at the Building Arts Forum Lecture series at the Guggenheim Museum in New York City. A collaborator with architect Renzo Piano and Ove Arup's Peter Rice on the Centre Pompidou in Paris, he was a good choice for the program's discussion on synthesis between architects and engineers. Equally, he was a prerequisite for a book on exposed structure. For Centre Pompidou, completed in 1977, he created a new architectural style in which all structural and mechanical systems are pulled to the outside of the building and exposed, leaving the inside free for the changing needs of the interior space.

THORNTON: What are your views on exposing structure?

ROGERS: Architecture is about giving order to construction and to the environment, and exposing structure is one of the ways to give order. We tend to put structure on the outside because we're looking for maximum flexibility of loft spaces. We believe that uses tend to have a much shorter life than buildings.

As we approach the twenty-first century, the one thing we're sure about is change, and, therefore, flexibility is paramount. We didn't just put the structure on the outside at Pompidou, we put all the services—staircases, electronics, and so forth—on the outside (Fig. 2.30). The inside is uninterrupted by any big concrete internal core.

This also puts the short-lifespan elements on the outside where they are accessible. Machinery such as air-conditioning equipment has a relatively short life. I should be very surprised if air-conditioning isn't completely revolutionized by the end of this century. Certainly we already know microcommunications didn't exist in buildings a decade ago. And now you can't even sell a building that doesn't take considerable care about leaving space for electronics. We don't know what the electronics will be, so we leave zones that are organized in the hierarchy of change.

I am very interested in what I call indeterminate architecture— buildings that can transform themselves. Therefore I am very interested in nonfinite, nonclassical forms. I believe the biggest change in our movement has been from the classical view of Vitruvius, where he would define an object perfect only when finished. Nothing could ever be added to it. Nothing could ever be taken away from it. I think that view is no longer possible. We're looking for something in which you can get harmony, but to which you can add pieces. It's like some forms of jazz. You play a regu-

Figure 2.30 (a,b) Putting all the structure and other services outside the skin leaves interiors unencumbered while making a strong, indeed quite controversial, architectural statement.
Centre Pompidou/Beaubourg, Paris, France; Architect: *Piano&Rogers*; Engineer: *Ove Arup & Partners*. (Photo **a**: *Martin Charles*, **b**: *Theodore Sherman.*)

lar beat but add elements on top of the beat. In architecture, you can see the basic beat is the structure. You have to have some organizing principle to the building, otherwise there is a lump.

If you look at Place Pompidou, the secondary elements are carefully designed, tremendously interchangeable. It doesn't matter if you have a red door, a black door, a shutter, or another corridor. We have outside corridors, streets to the sky, and sometimes you can plug those streets and layers on the outside of the building. It's very difficult to do that if the building has no strong rhythm, it won't hold that sort of change. We are searching for an aesthetic that will accept this form of change, and sometimes we use structure.

(b)

I have to be careful to say that not all our buildings have structure outside, although two or three of the most well known do (Figs. 2.31, and 2.32). But I can think of buildings where the structure is not on the outside, and I can think of climatic situations where it doesn't make sense. In sensitive environmental situations or in historical cities, it may not be logical. In certain situations, exposed structure goes against the program. I don't want to be dogmatic about it and say everyone should be doing the same thing.

THORNTON: We find that in the United States, even on a signature project, such as the United Airlines Terminal at O'Hare in Chicago, which we designed with Helmut Jahn, we have trouble with the cost of exposed structure. Clients resist it.

ROGERS: Maybe it has to do with the industrialization of the states. I think the States are the most limiting in my experience. You've got your industrial systems so well worked out to the nth degree of expenditure that it's quite difficult to do anything that is not absolutely standard. I suspect you have been forced to do more and more decoration—although decoration is of limited importance in my mind to architecture—because there is very little else you can do in that tight situation.

THORNTON: We believe it takes more hours of labor for architects and engineers to expose a structure because of the attention to detail required. The various systems must be so much more integrated. A lot of architects in the United States shy away from it because fees don't change whether or not a structure is exposed.

ROGERS: You're beginning to lose more and more control. And if you get a cold here, we get pneumonia out there—and we are getting it. There's a strong wind out there saying, "why do you need so much fee? We have American architects who will do it for half that." We say, "Yes, but the fee is related to what you are doing." How much can you afford to do? On a building like Lloyd's of London, the architect's office alone did about 5000 drawings. That's based on a fee of maybe 5 percent. In this country, you'd probably get half that amount of money. But your office would only do half that much work. I don't mean to oversimplify, but I am sure you understand the point.

We're not magicians, none of us are. When I have a client that can't afford the higher percentage, I can't do that much detail. But there's a problem. Slowly, we are being pushed into what I call a decorator position. We're getting more and more involved with beautiful boxes. And I don't think that has much to do with architecture. We should be worried about the whole planning implication. Should the building be there? As the developer's financial motive becomes more and more acute, I think architecture suffers.

Figure 2.31 Exposing structure and mechanical services benefits from attention to design, such as the careful detailing that is evident here. **Lloyd's of London, London, England;** Architect: *Richard Rogers Partnership*; Engineer: *Ove Arup & Partners.* (Photo: *Richard Bryant.*)

Figure 2.32 The central hall is dramatized by exterior exposed trusswork. **Lloyd's of London, London, England;** Architect: *Richard Rogers Partnership*; Engineer: *Ove Arup & Partners.* (Photo: *Richard Bryant/Arcaid.*)

BERNARD TSCHUMI

Bernard Tschumi offers an interesting perspective on the practice of architecture. A theorist—first a professor at the Architectural Association in London and now dean of the Columbia University School of Architecture—he also has an active architectural practice, stimulated by entering (and often winning) international competitions such as the Parc de la Villette completing construction, a Center for Arts and Media in the north of France, and four inhabited bridges in Lausanne, Switzerland. Other projects include the National Library in Paris, and Kansai Airport and the Japan Railways Kyoto Station development, both in Japan. Tschumi does not believe in exposing structure for its own sake, but he delights in working with engineers to "push the idea of building one step further" when and where appropriate as an element of a building or program. He finds encouragement in the current group of architectural students, who are fascinated with new technologies to assemble buildings and to model them.

THORNTON: How do you feel about the whole subject of when and where structure should be exposed or expressed?

TSCHUMI: If you go back in the history of architecture some 2000 years, it's shown that the great conceptual leaps coincide with technological and structural advances. Take the last 200 to 300 years, for example. Cast iron and railway stations suddenly brought a new type of architectural space. You see it in skyscrapers with the invention of the elevator. You see it more recently in the advance of glass technology or in computers. With computers, it's not the materials, but the ability to calculate extremely complex structures that has allowed buildings to go one step further.

In the same way that a mathematician tries to push the limits of the discipline to the next step, an architect has to understand what can be done through technology and structure. This is a starting point.

It doesn't quite make sense to see everything that you design or build as a new structural tour de force. Often you would be wasting a lot of energy and a lot of money by misdirecting it.

But architecture is the materialization of a concept, and concepts can be underlined or emphasized through the use of new technologies in your structural systems. A concept can be a lot of things. It quite often has to do with a program—whether it's a hotel, a swimming pool, a sports arena, a railway station, or an airport. And quite often contemporary programs are about a little bit of everything. Recently, we were involved with a competition for the City of Kyoto, where in the same complex we had a convention center, a hotel, a swimming pool, a series of dining rooms, a ballroom, and so on. We decided to treat the conventional areas of the project like hotel facilities, hotel rooms, office space, and whatever, in a conventional manner. Other parts which were more special—the great hall for the passengers to filter through, the swimming pool, the banquet hall, the so-called multimedia cinema, the electronic games, and all these things— were the most "active" parts of the program and were hence translated in the most "active" structural element. Not that I wanted a structural tour de force, that's not the idea, but to turn a key aspect of the project into an "image." We emphasized the specificity of this "program," of this "image," through a new approach to structure.

THORNTON: Can you point to other examples?

TSCHUMI: Yes. The Parc de la Villette, which is composed actually of points, lines, and surfaces (Fig. 2.33). The points are the red structures. The lines are a series of covered walkways, three-quarters of a mile in one direction, half a mile in the other direction, with a bridge over a canal. The idea here, because they were special elements, was that we would try to push certain aspects of contemporary structures as far as we could. The north-south walkway is actually a 1-kilometer-long beam with supports. The beam goes under a series of structural constraints that are very different from one another, and it is interesting to see how they are taken into account, exposed or hidden, in the design of the overall structure. That was the beginning of our relationship with Peter Rice who is, I think, one of the most extraordinary minds in terms of structures today. We had a lot of discussion on what is stable and what is not stable; what appears stable and what does not appear stable. At the time when so many so-called deconstructivist architects are interested in chaos and explosions of all sorts, it was very interesting to understand also how far you could push things structurally and at which moment everything would cease to make sense.

I'll give you a third example, which is now specifically about the relationship between an architect and an engineer on a project, in this case a competition for the New National Library in France. We were interested in the concept of circuits. The whole library was about circuits. Circuits of information, circuits of

Figure 2.33 This design includes walkways expressed as structural beams.
Parc de la Villette, Paris, France; Architect: *Bernard Tschumi Architects*; Engineer: *RFR*. (Photo: *J. M. Monthiers, courtesy of Bernard Tschumi Architects*.)

books, or people, etc. And one of them even was pushed to the extreme and became a running track. One idea was that the intellectual of the 20th century was also going to be the athlete of the 21st century. To support that running track, which was also the exhibition area required in the program, we started to suggest a structural system. But we weren't sure about it, so we sent it to Peter Rice and his team. They noted it and made a suggestion. It came back to us by fax and we said, "it doesn't look quite right—too boring." Once again, we really wanted to make the key part of the project an event. "It just looks too predictable." Fax it back to them. And then the thing comes back a couple of days later. And we say, "Hey, what's that? It's never going to stand up. It'll just collapse!" So a new set of faxing. There came a moment when they succeeded in making us nervous just as we tried to make them nervous. Then we got excited. And we wondered, "How do they do that? How does it work?" And the thing is quite extraordinary. It looks totally unstable but—yes—it works.

That thought brings me back to Kyoto Station. There's a beam that contains the tall and slender towers and holds all the activities—the swimming pool, the banquet hall, the multimedia theater, and so forth. It's a beam that's horizontal, a very large caisson that is about three stories high and 250 m long supported by seven towers. But the caisson is so large that you realize you don't need all the seven towers supporting it. So Peter at one moment said, on a couple of the towers you might not need to get down to the ground. Maybe they can stay suspended in the air (Fig. 2.34). The remark was great because it was saving considerable foundation costs over an existing subway line *and* it was reinforcing the architectural concept.

THORNTON: Like that old architectural form where you see what appears to be a pair of arches, but the middle column comes down and stops part way, so you think there's a column missing?

TSCHUMI: Could be something that way. It was great conceptually, and it was great in terms of actual functional problems. So suddenly a dialogue starts to develop. It's not only the dialogue about how you're going to make it stand up, but it's about how can you push an architectural idea to the next step. In other words, you're dealing with the advancement of the discipline. It's not anymore, "Let's design something and let's make it stand up," I mean the engineers. It's, "How can all of us push the idea of building one step further?" And I think that's a very, very exciting notion. And that's what we are trying to do.

THORNTON: That brings up an interesting point. You realize that to a structural engineer, when the structure is exposed, every detail is seen, and the efforts required to attain the structure are appreciated. But when the structure's covered up, say it's clad with stone, my perception is that no one sees it or appreciates the things we do. Let's say, for example, that we do a building with a large open space at the base.

Figure 2.34 Two of these tall thin towers stop above grade, carried by the horizontal "beam" building.
Kyoto Station Development Competition, Kyoto, Japan; Architect: *Bernard Tschumi Architects*; Engineer: *RFR*. (*Drawing courtesy of Bernard Tschumi Architects.*)

Figure 2.35 Exposed columns highlight a running track.
Paris Library Competition, Paris, France; Architect: *Bernard Tschumi Architects*; Engineer: *RFR. (Courtesy of Bernard Tschumi Architects.)*

Loads are transferred through all kinds of gymnastics, but no one notices or cares. Most architects don't have the reaction you had when Peter sent that fax back—"Now I'm nervous." Somehow when it's covered up they are not nervous. They trust us.

TSCHUMI: I want *not* to trust you. I want to push you to go to the next step so that you can also force me to go to my next step.

Maybe we should look at some of the models here. Where do I start? This is the Paris Library Competition where I mentioned the series of circuits, and the public is very much in the red area. Then at the top you have the exhibition ring right inside; on the top of the ring you have the running track. This ring is being held by those "skylons," those pylons slightly reminiscent of the 1950s at the time of the Festival of Britain (Fig.2.35).

THORNTON: When the average structural engineer looks at a column or a post, he or she likes to see it either vertical or symmetrically inclined or placed, but the skewed columns here can also work.

TSCHUMI: Exactly. But let's go back to the Kyoto Station, which we can see better on the model over there. Here are conventional functions, such as the hotel, the convention center, and the department store. Then in front you have that long horizontal beam with those very thin towers. They are only 15 feet wide, with some of them supporting to the ground, some of them still hanging in the air.

THORNTON: What sort of functional use is occurring inside there?

TSCHUMI: There is the multimedia theater, then the banquet hall, a conference hall with wide wedding rooms for the hotel, and the health club with a swimming pool. This is a level of restaurants for the department store and underneath are electronic games. In the towers intersecting programmatically you find some of the similar functions, but also here a series of bars—typical Japanese tiny, tiny bars. You know the floors are 15 by 30 feet, very small. Here we have a kimono museum which was required—again small rooms; here is an extension of some of the big conference halls, not very big, then tiny meeting rooms in the offices of the very slender tower. Here are waiting rooms for the wedding and flower displays, the health club, the aviary, the Emperor's waiting room, and so forth. They were all required by the program.

THORNTON: Will the structure actually be exposed?

TSCHUMI: It is exposed. This is the Kyoto Station building where you have that long truss or caisson, the hanging glass curtain, and the slender towers. Here is one of the towers that does not go to the ground, it floats in the air...

THORNTON: Is there vertical transportation in here?

TSCHUMI: Yes, of course. In two cases the elevators actually go down, but the structure does not. So the whole thing was absolutely stunning. We didn't use any diagonals ....

THORNTON: More of a vierendeel-type truss.

TSCHUMI: But the vierendeel system applies to the towers and—you have some details here—you have both the horizontal and the vertical (Fig. 2.36). And then even the use of glass in a climate where you have earthquakes—the idea was again very much developed with Peter Rice—you have a hanging curtain of glass which would have its flexibility. So I come back to my very first statement: The concept of the project was to say there are certain parts which are conventional, namely, the hotel rooms and the office space. But where the project was exciting, that's where we put all the

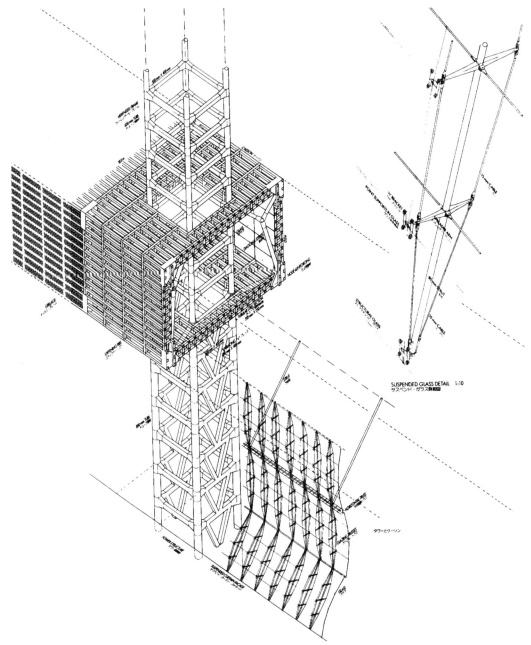

Figure 2.36 Along with the horizontal and vertical building elements, the glazing system is an intricate structure in its own right.
Kyoto Station Development Competition, Kyoto, Japan; Architect: *Bernard Tschumi Architects*; Engineer: *RFR*. (Structural axonometric: *courtesy of Bernard Tschumi Architects*.)

structural excitement as well. So suddenly this becomes the image, the scene which people will remember, right?

THORNTON: Right. This would be an excellent example for the book, especially that section through the glass wall.

TSCHUMI: It is actually quite exciting to do a design like this, but then the client was worried about the Buddhist monks and the preservationists, and chose one of the more conservative projects by the seven invited architects.

THORNTON: How about other buildings?

TSCHUMI: Another example is here, which I am sure you have seen; it has just been built. It's a small gallery, a video art gallery in the north of Holland, and we were interested to use glass in the structural sense (Fig. 2.37). That is, the beams, the supports, the fins are all made of glass. In other words, the whole thing is a complete self-supporting structure by simply using clips. Common technology now is not used in this way. We are pushing it a little further, and then using certain devices because of the fact that there is a disorientation. If you tip it a little bit, when you walk up you don't quite find your balance because your body is not used to walking at an angle. Then, on top, where the glass is transparent and there is a reflection, you get a strange feeling.

THORNTON: That's an excellent example of structural glazing. I don't know of many other examples except for the Pilkington system for windows (Chap. 6).

TSCHUMI: The gallery is quite fantastic. At night, and even during the day, it is a very interesting thing. You sort of feel your way through it, it's like a ship. At night, what happens is that reflections of the video monitors all over the place mean that you have no sense of the way space would begin. It is like a hall of mirrors—very fascinating.

THORNTON: What sort of joint is between the glass components—a butt joint?

TSCHUMI: A sealant, just so there's no leakage of water. This is an open grating, and there's a door at both sides. There's enough air circulating through so there's no condensation.

THORNTON: I asked Richard Rogers and Peter Rice this question: Do you see more receptivity in Europe for these kinds of structures than in the United States? There's such an emphasis on the bottom line here.

TSCHUMI: In the United States in most of the 1970s and 1980s the fascination with the past, with historicism, with trying to pretend that we've gone back to the 18th century has been a force not terribly conducive to developing anything new. And this is a

Figure 2.37 Glazing provides all the structure of this video store.
Glass Video Gallery, Gröningen, Netherlands; Architect: *Bernard Tschumi Architects*; (Photo: *vander Vlugt & Claus, courtesy of Bernard Tschumi Architects.*)

little bit of an unfortunate situation, which I think we are now just coming out of. Many of the younger generation of architects are absolutely fascinated with new designs and technology. I see it at Columbia. I'm sure you see it also. The younger generation is fascinated by new ways to put buildings together, new materials, not only the actual structural technologies, but also having to do with the production of simulated images, virtual reality, and all that. This is having an effect on the new architectural sensibility, which is, of course, totally different from doing neo-Palladian or neo-Georgian architecture. At the same time we're still at the end of that sort of very conservative backlash of the 1970s and 1980s. My students are fascinated by the structures that were done in the 1950s.

THORNTON: What about litigation and liability? Do you think that sort of dampens enthusiasm in the United States to do new things? I find that in our own business in the 1960s and early 1970s we were more daring than we are today because of the litigation.

TSCHUMI: You think it's because of litigation?

THORNTON: Yes. We would always share with our clients what we were doing that was revolutionary or different, and I find that when we try that today, our clients usually say don't do it.

Look at the dorm at Columbia with all the tile falling off it. That was not a revolutionary use of the material, but it was different for this area.

TSCHUMI: Everyone was upset about that. But no, I don't think the litigation aspect is enough reason to avoid trying new things, because that applies to every aspect of architectural practice. At the same time when you do try something new, you are taking a certain amount of chances, and you need those clients who want to go along with it. I do think that you structural engineers are also responsible people and generally know what you're doing. And architects also by and large are trying to do their jobs seriously.

3 *Conversations With Engineers*

INTRODUCTION

In this book, "engineers" refers to structural engineers who design building frames. This is not to ignore the importance of mechanical, electrical, and plumbing engineers to building design—without their contributions, modern buildings would be inaccessible, uninhabitable shells—but their main impact on exposed structure is finding ways to route services unobtrusively (which can be an art in itself).

While the public may view engineers as silent supports for the machinery of the modern world, the people in these conversations are quite vocal about their ideas and their experiences in the process of building design. The secret to get them talking? We asked about their "children"—their favorite projects. Then, as structural engineers speaking with other structural engineers, our conversations inevitably turned to "shop talk" about fees, schedules, and relationships with architects.

We have tried to let the tenor of each conversation and the personality of each speaker come through the text. If you are an engineer also, we think you will find that many points will bring a flash of recognition—"yes, that's how it is!" If you are not, welcome to our world!

ELI COHEN

Eli Cohen is a principal of the Chicago firm of Cohen-Barreto-Marchertas Inc. (CBM Engineers). Eli and his firm have established the reputation of being the premier structural engineering firm in Chicago. Special to Chicago is the extensive use of concrete, not only as structure, but as the entire facade. The authors were intrigued by the substantial use of exposed concrete in Chicago, although it is not prevalent in other major cities in the United States. During a visit to Chicago in preparation for a joint presentation for a project with CBM, Charles Thornton had an opportunity to sit down with Eli and review some of the background of the prevalent use of exposed concrete in Chicago.

THORNTON: One of the things we have noticed is that exposed concrete structures in Chicago are really special. Nowhere else in the country do people put up so many exposed concrete facade buildings. In New York City you find two or three buildings with exposed concrete and in Chicago you find dozens. I think you've designed most of them. I am curious; to what do you attribute this extensive use of exposed concrete (Fig. 3.1) ?

COHEN: I would say, one, the quality of the concrete in Chicago is excellent. Control of the material, getting the right strength and the right consistency. Two, local contractors are very efficient in doing their work, which helps too, especially considering that the climate is not the most ideal here for concrete exposure. But looking back at most buildings in Chicago, they are holding up pretty well.

THORNTON: When we propose an exposed concrete structure in New York City, we look at the relative cost of the concrete. If we have a plain reinforced-concrete structure that gets covered up, let's say the total unit price might run $300 a cubic yard including everything. If we price it as an exposed concrete job, the price goes up to $600 or $700 a cubic yard. What happens in Chicago when you do an exposed concrete job? What happens to the price compared to putting a facade over it?

COHEN: It adds a certain amount to it, but I don't think it goes up in the same proportion. Once you have quality forms available and if you use the same form as you go up, at the moment I can't tell you the percentage it goes up, but I would say it's quite reasonable, it's not double. It depends on the forms, how many times you can reuse a form. You know as the building goes higher up, the forms begin to deteriorate and so does the concrete finish. But if you are high up, nobody can see it.

THORNTON: That's a good point. Most of the formwork in New York City is built up with 2 by 4s, 4 by 4s, and 4- by 8-foot sheets of plywood—wood forms. What you are saying is that in Chicago you use engineered forms, steel forms, plastic lining, and things like that.

COHEN: Steel forms, or fiberglass forms, and the design of the forms become critical to take care of the varying conditions. You can't have sharp corners. In normal practice if you take care of that, you get excellent results.

THORNTON: I feel that the unions have something to do with the problem in New York City. I think that if you try to bring preengineered, prefabricated forms into New York City, you will have a problem with the unions because it will reduce field work. We still bend every rebar in the field at the job site. Nothing can be bent in the shop. What's the union situation in Chicago?

COHEN: From my understanding you get the forms shipped to the job site and put them together there. Also as you said, the reinforcing is part of it too. That's another handicap you have in New York. You have to bend all the reinforcing bars on the job site, and that goes hand in hand with exposed concrete. If you get everything prefabricated and shipped to the job or put together on the job, that's where economy and quality come in.

THORNTON: You mentioned a minute ago that the climate in Chicago is cold. When I came here this morning, it was 50 degrees in New York City and 16 degrees here. When you don't have insulation on the outside of the structure and you have exposed concrete columns and spandrels on a 40- to 50-story building, how do you compensate for the differential behavior between the outside columns and the inside columns or the core (Fig. 3.2)?

COHEN: I think when you go to the very tall buildings, beyond 40 stories, you have to watch more critically. To account for the shortening due to the temperature drop, that is, shortening of exterior columns in relation to interior columns, you get quite a differential. We have developed a method of taking care of the stresses which develop in the upper slabs as the columns shorten due to the temperature. Also, you have to watch that the columns are not overinsulated from the inside. You want to expose as small a portion of the face of the column as possible to outside temperatures and have much of the column inside. Also you don't want to insulate the columns excessively. You can balance the heat from the inside and the outside so that it cuts down on the shortening of the column due to the colder exterior temperature. Usually, there's a certain dialogue between engineer and architect in finding a compromise on that.

THORNTON: Does the architect provide details of the partitions which run perpendicular to the outside so that there's a slip joint? What do they do to eliminate the stress in the partitions?

COHEN: The architect details the partitions which butt against the exterior columns, so they are not really butting against the concrete. Keep a certain gap in between, and that often takes care of it. In addition to that you have to take care of the way the partition hits the ceiling and the floors so that it allows for the column shortening. You will get a certain amount of deformation of the floor slab, and if you don't take care of it, it will just crack the partition or damage it.

THORNTON: Have you ever gone out into the field on any of the buildings a few years after completion to measure how much the outside column elongates in the summer or shortens in the wintertime? Has anybody ever gone back and done that?

COHEN: I believe the Portland Cement Association did some evaluation of it. There was a paper written on it by Mark Fintel. He went through and did quite a bit of checking on this subject. When it comes down to the elongation of the interior column, we don't really pay too much attention, because the interior column temperature is usually about 70 degrees (F). The exterior column could go up to 90 degrees or so, but the shortening due to cold weather could become critical and have a sort of bowing effect on the column.

THORNTON: What kind of surface treatments are done after the concrete is stripped? Do they put any coatings on these things or is it just exposed concrete?

COHEN: The earlier buildings were just exposed concrete. Another thing you have to watch out for in some buildings is calcium chloride. Our firm has never allowed calcium chloride in any columns where material is exposed to weather. A number of buildings which were done in Chicago used calcium chloride, and it had a very negative effect on the spalling and delamination of the concrete. As far as the sealing of the concrete, there are different products on the market, but at the moment when we have exposed concrete, we call for breathing surfaces and sealants that don't seal it off completely, but allow it to breathe.

THORNTON: What's the tallest exposed concrete project in Chicago that you've done?

COHEN: We did the Newberry House, which I believe was a 56-story exposed concrete building (Fig. 3.1). River Plaza was about the same height (Fig. 3.2). We did a number of office buildings, such as 180 North LaSalle—these were about 40-story office buildings (Fig. 3.3).

Figure 3.1 The visible grid of this facade is an exposed structural concrete frame. **Newberry House, Chicago, Ill.**; Architect: *Gordon & Levin*; Engineer: *Cohen-Barreto-Marchertas, Inc.* (Photo: *CBM Engineers/Chicago.*)

Figure 3.2 Structure here extends well outside the plane of glazing—significant in this cold climate.
River Plaza, Chicago, Ill.; Architect: *Ezra Gordon, Jack Levin & Associates*; Engineer: *Cohen-Barreto-Marchertas, Inc.* (Photo: *CBM Engineers/Chicago*.)

Figure 3.3 This flush concrete frame creates an impression of stone panels.
180 North LaSalle, Chicago, Ill.; Architect: *Harry Weese*; Engineer: *Cohen-Barreto-Marchertas, Inc.* (Photo: *CBM Engineers/Chicago.*)

THORNTON: What's been the experience? After making the decision to use exposed concrete instead of putting a facade on it, are owners happy with the buildings?

COHEN: I hope so. It's a question of economics. Recently, we did the 60-story North Pier building. We played around with exposed concrete and finally ended up using a precast cladding because it was a requirement of the site. We couldn't have an exposed concrete building there.

THORNTON: Does that mean that there are some negative images of exposed concrete? Do certain people in Chicago perceive that an exposed concrete building is inferior or of less quality than one that has a cladding?

COHEN: No, I think it was part of the North Pier—the requirements of that whole site north of the river. All of the buildings have to be clad. It's part of their charter.

THORNTON: If you do an exposed concrete project versus a non-exposed concrete project, do you spend more hours? Do you have to be more meticulous?

COHEN: You have to pay more attention to some of the details and you have to work hand in hand with the architect. If the architect wants to express the structure, you have to watch out that structurally it works and that also the effect of the climate wouldn't be too negative. Overall the design doesn't take much more than having a clad structure.

THORNTON: After the Arab oil embargo in the early 1970s, when the state energy codes became a factor, did that create any problems with exposed concrete structures? More heat loss? Have you run into situations where the architect and the mechanical engineer thought that exposing concrete made it harder to meet a state energy code?

COHEN: It's a little bit harder. Again, when we talk about insulation, we reduce the insulation over the columns, not over the spandrel beams. We still feel the spandrel beams have to be insulated. It has an effect, but overall. If you've noticed, it's not only the concrete exposed, the windows are smaller and other things happen, so the concrete is not the only issue there.

THORNTON: Does anybody ever complain that when you touch the inside of the column on a cold winter day it's cold? Any adverse problems with condensation or any kind of sweating?

COHEN: Usually the columns are furred out, drywall, $\frac{5}{8}$ inch or so, but there's no insulation. Just keep a gap between. I imagine some people may be unhappy, but we haven't gotten any complaints.

VINCENT DESIMONE

The colorful Vincent DeSimone, chief executive officer of the New York City–based structural engineering firm DeSimone, Chaplin & Dobryn, was an excellent choice to talk about not only the technical but also the political aspects of exposing structure. He provides a lively description of the give-and-take, the demands and demurring among the architect, engineer, and, of course, the owner, that eventually results in a structure's being exposed or expressed. He points to the exposed linear steel trusses of the Niagara Winter Garden and the exposed concrete of One Biscayne Tower in Miami as the most significant exposed structural designs of his career.

THORNTON: What are the advantages of exposed, expressed structure? Disadvantages?

DESIMONE: Historically, you know, all structures were exposed. There was no dichotomy between architects and engineers. If you look at everything up to, I would say, the mid-19th century, there were bearing walls and masonry which created the great piazzas of the world. That's all structure. I think it was the Industrial Revolution that created the schism between the architect and the engineer. Modern materials, starting off with cast iron and going into mild-grade steel and then reinforced concrete, created a distinct split. You've got a building like the Eiffel Tower that's exposed structure. That, in my opinion, is probably the greatest exposed structure in the world. But it seems that architecture goes through these phases. It comes in 10- or 20-year cycles—people suddenly discover structure. They say, "Why can't the structure be left exposed as part of the building?" Then, for instance, we get exposed concrete. But concrete is ugly looking, so they say, "Why don't we do something to it?"

What you wind up with is a classification of architect-engineer, or engineers that become like architects. For example, you have Pier Luigi Nervi; his work is all exposed because the architecture comes from the structure. His concepts of space and the things that he's doing are structurally oriented. Take that marvelous roof that he has on the Mill Building, I believe it's in Milan, where you actually see the traceries of the stresses that occur around the columns and the columns are fluted and they come up...it could be the Fiat plant. You see these things, I believe, because he's an architect-engineer. Another example is Kenzo Tange. With Tange, everything is very, very strong. It's very structural. Tange is a structural engineer first, and he evolves as an architect from that. We've watched lots of great architects go through this transformation.

TUCHMAN: But you've worked often with Michael Graves. There's an architect for whom structure is completely subservient to the architectural goal.

DESIMONE: But even he's gone through a period in which structure was preeminent. Look at the Cincinnati Orchestra outdoor theater at Riverbend. That's pure structure (Fig. 3.4). Of course, there are little things all made up; there are towers and turrets. If you looked at the original sketches we did, Graves actually wanted the roof to be suspended by chains that came off the towers so it would really look medieval. But he went through a particular period somewhere in his mind of leaving the structure exposed—maybe because of its requirement for economy, maybe because of the fact that it was open air—and he wanted to express the openness of it all.

I think a lot of the exposed structures that we get involved with are due to the quality or type of building. For example, you've worked with Cesar Pelli. His designs are straightforward, but when he gets to open areas for the public, like the Niagara Winter Garden, they're complete and total exposed structure.

TUCHMAN: Was there more of this going on in the 1960s and 1970s than there is now?

DESIMONE: I think more than a tête-a-tête occurred in that period between strong structural engineers and very good architects. Architects were looking for something logical, but their designs frequently are not measured by logic. They're measured by taste, and, as you know, there are 50 elegant, wonderful architects who all have different styles. What happened is that they would come to an engineer who had a good reputation for being a strong engineer, and their inclination was to try to get something from him that would be exciting from an architectural point of view. For example, we worked with Sert and Jackson on the Harvard Science Building in 1968. We knew there was going to be a lot of vertical material, and we were looking at ways to pre-manufacture building components. That led to channel shapes and double tees, which led to trying to use the space between the verticals as ducts, and so forth. It's not fashionable today, but when we were working on that building, the ultimate goal was to get a building that would be totally and completely integrated, one in which every component would

Figure 3.4 Openwork support towers for this theater roof are both industrial and airy in nature.
Riverbend Music Center, Cincinnati, Ohio.; Architect: *Michael Graves Architects*; Engineer: *DeSimone, Chaplin & Associates*. (Photo: *DeSimone, Chaplin & Dobryn Consulting Engineers*.)

double its service for everything else. So a piece of structure became both a piece of architecture and a concrete duct.

THORNTON: Did it prove to be cost-effective?

DESIMONE: No. My opinion is that in most instances the tendency is to take the structure and rape it. But a beam is not a duct, and what happens is, you begin to forget that you need a way out of the vertical, so you end up chopping holes in the vertical legs and tees, which makes it impossible for them to stand. Or you wind up realizing that as the air is pouring through the concrete, it's picking up dust so that you're contaminating your entire air filter system. The whole thing gets crazy. In other words, the simplistic notion that one structural element could substitute for all the trades was simplistic and didn't work. It created problems in the field, and it created buildings that looked like industrial plants. I think the pendulum has swung away from exposed structure.

THORNTON: How do you explain the fact that in Chicago there are all these fabulously exposed, painted concrete jobs, but in New York City everything, to the best of my knowledge, is clad with something else?

DESIMONE: I absolutely agree. In New York we're suspicious of exposed concrete. I think what happened was that a tremendous amount of work was done in Chicago to overcome what happens to the interior of the building when you have a column that is semiexposed. When you've got a column that is exposed to 0 degrees on the outside and 76 degrees on the interior, something is going to happen to that structure. New York engineers have avoided the problem by simply saying, "It's unmanageable. I don't want a building that moves." But in Chicago they went ahead and solved the problem. They said, "Look, when you have the interior partitions attached to the exterior walls, set them into a track that allows 1 ½ inches of movement." They introduced thermal isobar types of calculations to determine the average movement and then managed to accommodate all that stuff.

I also think that part of the problem in New York is a bunch of cheap owners. What they did was to say, "Hey, listen, that's concrete. Why don't you put something on it to make it look good?" We wound up taking structural concrete specs and, in the beginning, putting a form liner in the spec. I mean, that was our big concession to its being exposed. But exposed architectural concrete is just not exposed structural concrete with a little cosmetic work. It's a whole different kind of thing. It's making forms that go by the pour so that the next set of forms doesn't get offset. I think Chicago was prepared to take this whole thing and spend the money to do it right, and they have achieved good results.

The validity of exposed structure exists, but you can't pack too many things into the structure. If it's done elegantly, the structure has to be all of what you have. Structural steel that doesn't have to be fireproofed, space frames—those things cry to be left open. They are so elegant structurally that architects should be shot if they put in hung ceilings.

TUCHMAN: What is the most exciting exposed structure you have ever been involved with?

DESIMONE: I would say that the Niagara Winter Garden is the most elegant structure that this office has designed. It was done by Gruen Associates, but the design was Cesar Pelli's. Essentially, it was a low-budget project, part of an urban renewal. What we really wanted was a glass enclosure so that there would be plants and landscaping year round and people could come to enjoy the area, even in Buffalo's harsh winters (Fig. 3.5).

It turned out to be an incredible structural problem because the structure was exposed and, to a large extent, we were supporting everything from about five large concrete columns. Although it looks like a space frame, it isn't. It's actually a series of linear trusses—large trusses in one direction and smaller trusses on a sloping face creating a chimney effect on the other side. Our biggest problem was the fact that, once again, when you expose structure, you're no longer just fooling around with wind, dead loads, and live loads. You now have to think of the thermal consequences. Because we were at the end of cantilevers and the sides of the Winter Garden themselves were sheer glass, our concern was how much the roof would move and where we should put the expansion joint between the roof and glass sides. If the expansion joint was on the roof and the glass was supported from the lowest level, then the joint would constantly be moving as the roof trusses changed shape due to thermal movement. The architect wanted a crisp roof-wall corner, didn't want to see a parapet, and didn't want to do a flyby where the sides rose above the roof. So what we did with that entire sheet of four sides of glass was to hang the mullions from the roof trusses and run a track along the lowest level. The mullions themselves were set into a vertical guide, and as the building responded to the temperature and the roof cantilever started to rise, it literally lifted the mullions up with the glass. The mullions were set or encapsulated in guides that allowed vertical movement, but not horizontal. And I was able to get a joint between the glass and the mezzanine.

THORNTON: This brings up an interesting point. One prominent structural engineer we interviewed took exception to my position, which happens to be the same as yours, that when you expose a structure, you have to take into account extra phenomena such as thermal, creep, and shrinkage. He pointed out that we have to do that anyway in all structures for the construction condition, so it really is no extra work at all.

DESIMONE: That's nonsense. The fact of the matter is that when we go ahead and clad the building, only under the most critical con-

Figure 3.5 Linear trusses create a space frame effect at this "greenhouse for people." Thermal movements at breaks in plane were given special attention. **Niagara Winter Garden, Niagara Falls, N.Y.**; Architect: *Gruen Associates/Cesar Pelli & Associates*; Engineer: *DeSimone, Chaplin & Associates*. (Photo: *DeSimone, Chaplin & Dobryn Consulting Engineers*.)

struction conditions are you going to take thermal stresses into account. An exposed structure has to go through 12 months a year of cyclical thermal changes, and the big problem is that certain parts of it are heated and certain parts are not. For example, the top chords in the Niagara Winter Garden are at a totally different temperature than the bottom chords, even though they are within the same glass structure. So in addition to having just straight linear movements and expansion, we began to get thermocouples simply because of the difference in temperature from top to bottom.

TUCHMAN: Other examples?

DESIMONE: I think my best exposed concrete building is One Biscayne Tower in Miami. It was designed by the Cuban architect Enrique Gutierrez, and he really wanted to show the guts of the building. He wanted the building to be it, and in between the columns he wanted to place a window wall, and that's exactly what he did. The building is very, very powerful. It has exposed concrete running up the full 500 feet of its height, with sidewalls acting as vierendeels to stiffen against hurricanes (Fig. 3.6).

But you've also got to understand something else. When you're dealing with exposed concrete, sometimes the architect says that no concrete shall be repaired under any circumstance. What he's really trying to do is tell the contractor that he can't afford any mistakes. I hate those specs. I think a lot of exposed buildings get ruined simply because of a lack of reality of what can be expected. That reminds me of what Philip Johnson did at the New York State Pavilion at the New York World's Fair in 1964. We thought the contractor had blown it completely. We were slip forming 16 towers 80 feet high and 12 feet in diameter. The mixed concrete was not setting at the proper rate, so that as the slip form was going up, we were dragging a lot of aggregate up with it. They were supposed to be slick columns, as slick as if they had a steel finish. Instead, they wound up being these organic, rough concrete shapes. I know the contractor was beside himself, and Johnson came down and looked at them. He said, "I love them. They absolutely look like they were thrust out of the earth with all its roughness." And I thought that was wonderful. Here was a guy who had his concrete finishes blown apart, and he was still able to accept what he was given and find a virtue in it. It was very practical.

THORNTON: I'd like to interject a story about a parking structure with an exposed spandrel. To get the project approved by a local borough and the local borough's architects, we agreed to let the

Figure 3.6 Exposed concrete fin columns stiffen this building against hurricane winds. **One Biscayne Tower, Miami, Fla.;** Architect: *Enrique H. Gutierrez, AIA*; Engineer: *DeSimone, Chaplin & Associates.* (Photo: *DeSimone, Chaplin & Dobryn Consulting Engineers.*)

borough's architect select the exterior precast panel finish. It really didn't make any difference to us. A panel is a panel. So we took him down to the precasting yard where we had lined up about 10 to 15 sample panels. He looked at the 10 to 15 that were lined up, and then he noticed one over in the corner. He said, "I want that one." It was the back of a panel.

DESIMONE: You're absolutely right. But there were some wonderful things that we did with exposed concrete in garages. If you look at Paul Rudolph's concrete garage in New Haven, it has aged well. But you have to remember when you expose a structure, the concrete by itself is not watertight.

TUCHMAN: So what did you do on the building in New Haven?

DESIMONE: Painted it. They have these marvelous, wonderful 20-year Tnemec paints.

THORNTON: Does the paint waterproof it?

DESIMONE: The paint waterproofs it, and it's a vapor barrier, but it breathes. So if you get trapped moisture behind it for any reason, it dispels it out the surface.

TUCHMAN: Did you paint the Miami building?

DESIMONE: We used a cementitious paint, which left the form joints visible, left the little tie holes visible—it really looks like gray plated armor. That building was conceived as an exposed concrete building, but I might add that it happens to be in a very forgiving environment, except for moisture. Oh, you know, it gets down to 40 degrees sometimes. And this stuff was sealed, big time.

THORNTON: But again, the moisture is not the culprit nearly as much when you don't have freezing.

DESIMONE: Right, freeze-thaw is the culprit.

I'd like to tell you a case in which the argument for exposed structure wasn't made by the engineer or the architect; it was made by the developer. It was when we did 17 State Street in New York City. The developer, Mel Kaufman, came in one day and looked at this cross-braced bundle of columns. (I had to bundle them because I couldn't otherwise get enough stability out of the building.) He told Bob Sobel, the architect, "You're not going to cover that, are you?" Sobel said, "Sure, of course I'm going to cover it." And Kaufman said, "No way. I love it. Do something with it." So now when you walk into the lobby of 17 State Street, there is a glass-enclosed elevator shaft. You actually see the elevator with the cable, the cross braces, everything (Fig. 3.7). He's the kind of developer who says the structure is probably the most important part of the building.

Figure 3.7 The developer liked this cross bracing so much he insisted that it be left exposed. Glass elevator shafts add to the effect.
17 State Street, New York, N.Y.; Architect: *Emery Roth & Sons*; Engineer: *DeSimone, Chaplin & Associates.* (Photos: *Len Joseph.*)

THORNTON: What was Richard Roth's and Bob Sobel's first reaction to exposing all that stuff?

DESIMONE: They didn't care for it one bit. Not only that, we exposed the roof truss. This job was one of the screwiest projects I was ever involved with.

THORNTON: Wasn't Ruderman's firm involved?

DESIMONE: Ruderman's firm had been involved in the preliminary engineering. But then Mel Kaufman decided to have an engineering competition. He sat there and asked, "How much do you think you can design this building for? How many pounds of steel?" I told him I would take it over the weekend and work it out. I made a concrete building out of it, because like you, I believe you can buy stiffness cheaper with concrete than with steel. I came back with it after the weekend, and he said, "No, no. I don't want concrete. I want steel." I had one night to go back over it. So I threw some numbers together and the next day told him that I could make it for 24 pounds of steel per square foot. He looked at me and said, "You're crazy." I asked him why. He said that he had just got an estimate of 38 pounds per square foot and financially could not make the building work. I said, "Listen. I don't know what you did with your other engineer, but I can do it for 24." He agreed that if I could do it for 24, we had the job. Now I was stuck with 24 pounds, and I can't even design well, I only conceptualize. It's my partner who's got to do all the work. So I came in and said to Carlos Dobryn, "Amigo, we have it. I got 17 State Street; that's the good news. The bad news is, it can't be more than 24 pounds. Kaufman will kill me if I'm wrong. He'll sue me for every pound above that."

We started to make up structural systems. We bundled the core; it was still too loose. Then I took all the perimeter spandrels, and I actually was able to get a steel guy to come in and give me deeper members for the same weight and for the same price per pound as rolled sections. So I wound up taking the bundled tubes, I wound up with the perimeter deep frames, and in the end we were still missing stiffness, we were at the height divided by 380. I still couldn't do it. Then we put a hat truss on top of the building, and that stiffened it back up to height divided by 500, close enough. About that time Mel came in and said, "I love that, I want to leave the bundled tubes." Bob Sobel, the architect, said, "I hate it." Mel said, "I'm the owner, I want to leave it." Bob said, "I don't want anything to do with this project, I'm finished; I'm tired of your designing the building. It's all over." Mel said, "Good, you're fired." But naturally they all stayed on the job.

THORNTON: I do this all the time. If you want to win a job, a commercial office building, you've got to commit to bring it in at a number that is maybe a reach. I did one in Pittsburgh in which I came back and said, "17 pounds—if you go over 17 pounds, I'll kill you."

DESIMONE: And you know what you did? You forced everybody to reach, to come up with nice technology and a whole bunch of structural tricks that all in all brought the price down.

THORNTON: We're a little bit off exposed structure.

DESIMONE: No, we're not because you see what happens...How much of this is exposed structure, and how much is not? This is not what I'd call exposed structure, but it is an exposed structural form. That's where the structure has been manipulated cosmetically to make it more appealing. There are varying degrees of exposed structure, I guess that's what I am really getting at; there is exposed structure like the Winter Garden and exposed structural form like 17 State Street.

EUGENE J. FASULLO

We sought out Eugene J. Fasullo, chief engineer of the Port Authority of New York and New Jersey, to get the input of an engineer working for a major owner. Fasullo spoke from his office on the 72d floor of the World Trade Center of the long-term view of the public owner of projects such as that very building. Looking at his expansive view of New York harbor and New Jersey, Fasullo noted that the structural solution of the 107-story building had to mesh with the architectural solution, and both were strongly driven by economics and concern for the users. Fasullo started his career at Port Authority in 1958, working on the addition of the lower level of the George Washington Bridge. In 1965 he had direct responsibility for the structural design work at John F. Kennedy International, LaGuardia, and Newark International airports. He conceived the structural system for the unique pile-supported prestressed-concrete runway extension at LaGuardia Airport, which was completed in 1966, and was responsible for the structural design of the concrete shell roofs of the terminal buildings at Newark Airport.

FASULLO: The unique perspective I bring to a discussion of exposed structure is that of a public owner. We are not only the architect and the engineer in many of our facilities, but also the owner; it is our money. Dealing with the total system of construction allows us to establish different priorities. Our concern is the total economic cost of a project, and in many cases we are willing to spend more on what would be consulting fees—the engineering effort, the architectural effort, model building, and renderings—because of the public feature of our work and because we are trying to save on the capital construction costs rather than just save by minimizing the engineering activities. Engineering costs may run 10 percent of the project; we wouldn't mind if they ran 15 percent of the project as long as we were able to reduce the capital project by something like 25 percent. That's where the real money is. So we have a different goal as an organization than either an architect or an engineer would have. Although architects or engineers have these global goals as well, they also have their own proprietary interests. Also, they represent a lot of owners who don't have the background and sophistication in the construction process to realize that if you invest some money in the professional design process, you have a great opportunity to save significantly on the capital side.

TOMASETTI: You're not talking about saving on the life cycle; you're talking about the initial cost.

FASULLO: Since we do maintain and operate these facilities, we also strongly consider the operating costs. As a self-supporting public agency, the revenue stream is very important to us. We are concerned about rental space, the same as a private owner would be. So we optimize not only in the initial cost (which is certainly a focus of the operating cost) but also in the maintenance cost, the policing cost, the safety issues, the risk analysis, and the liability exposure for those public uses of our facilities. Our professional work is within this very broad context.

TOMASETTI: We hope to get at the problems involved in using exposed structure—when you should use it to contribute to the building and how you should do it.

FASULLO: Our philosophy is that good design normally reflects the structural requirements of a project. I think the expressions of architecture and structure are almost automatic by-products of good design.

For example, the World Trade Center, which was the world's tallest building at the time of its construction, demanded a unique approach. A big challenge on a building like the World Trade Center is being able to carry wind forces effectively and economically. In response to this, a special concept was introduced that used a tubular design rather than a moment frame. To get as close as possible to pure tubular behavior and to minimize the secondary effects of bending of the exterior members, the World Trade Center's windows were made narrow and tall and its columns were very closely spaced. The result was an aesthetic approach by the architect in which the structural requirement was used as an architectural expression (Fig. 3.8).

In addition to providing a structure that was cost-effective, we also had to consider the physiological experience of the people using this high-rise building. As an owner concerned with renting the building, we didn't want a building that had any sense of discomfort. It was felt that if the windows were no wider than a person's shoulders, the occupants of the building would feel safe and secure and would not have a sense of being able to fall out.

We also had to consider fire-safety requirements, which precluded exposing the structural system itself. But the need for glazing and cladding was apparent. The result was that the structure is duplicated in architectural form, using only architectural materials and fireproof materials as protection.

In this example of the World Trade Center, the structural solution meshed with the architectural solution, and both were driven strongly by the economic requirements and the physiological experience of the people who would be using the building.

TOMASETTI: You personally had a lot of experience 20 or 30 years ago with totally exposed structure, such as hyperbolic paraboloids in Newark International Airport (Fig. 3.9) and weathering steel and concrete ramps at LaGuardia Airport. In your opinion, why isn't that happening any more?

FASULLO: I can't say that it's not happening at the Port Authority by any overt decision. At LaGuardia, where we have a very large parking garage for 3000 cars, the driving force was the fireproofing issue. At the time it was built, fireproofing was a requirement. However, we knew that technology was leading in the direction of not requiring that structural elements of parking garages be fireproofed. So while today the garage conforms to the code, it did not at the time we designed it.

TOMASETTI: So you were actually a pioneer in a sense.

FASULLO: Design pushes code writing. Code officials react to what the profession is demanding. In the case of the LaGuardia parking garage, there was a very expensive foundation requirement. LaGuardia is built on an old dump site, and because of the unique nature of the soils, the foundations are typically on piles and the piles are very long, over 100 feet long. Therefore lightness was

Figure 3.8 The closely spaced columns seen here form a perimeter "tube frame" to resist wind loads and define the architecture.
World Trade Center 1 and 2, New York, N.Y.; Architect: *Minoru Yamasaki*; Engineer: *Skilling Helle Christiansen & Robertson.* (Photo: *Jerry Rosen, Port Authority of New York and New Jersey.*)

Figure 3.9 Four hyperbolic paraboloid shells meet at a column to form the roof module here. Building curvature causes shell sizes to vary.
Newark International Airport, Newark, N.J.; Architect and Engineer: *Port Authority of NY and NJ.* (Photo: *Jerry Rosen, Port Authority of New York and New Jersey.*)

important. We did consider exposed precast concrete, but the economy of using steel was very significant.

The next issue was the architectural expression we could achieve without adding cladding as a separate element. We don't really think that what I call "hang-on architecture" is the way to go. Architecture, in our opinion, should be a natural evolution of the structural requirements. There are two advantages to this. First, there's a certain basic directness of architectural expression, and second, there's a cost saving by eliminating the need to apply architectural finishes, which are usually very expensive.

The only premium we paid was for our choice of weathering steel. We had to do model studies to determine the actual structural behavior of the exterior wall which, while being the architectural expression, was also a vierendeel truss composed of pre-

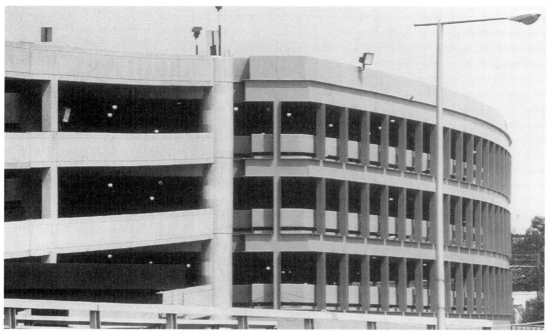

Figure 3.10 This parking garage uses exposed weathering-steel vierendeel trusses to span between supports. Architectural design dictated elimination of outer flanges at beam-column joints.
LaGuardia Airport Parking Deck, New York, N.Y.; Architect and Engineer: *Port Authority of New York and New Jersey.* (Photo: *Jerry Rosen, Port Authority of New York and New Jersey.*)

fabricated trees bolted together in the field (Fig. 3.10). So there is an architectural cost associated with using the structure architecturally, but when compared to that of applying another architectural material, it's really much more economical.

In using exposed weathering steel you've got to be careful that it behaves as one would hope, that rusting does stop. You must allow the steel to dry after being wet. If it stays wet, it will continue to oxidize at approximately the same rate as normal steel. We had the experience of certain joints not drying between rainfalls. The rust accumulation created enough force in the joints that over the years we actually broke a number of high-strength bolts at these connections because of the continued corrosion and expansion of the rust. Another factor with using weathering steel is that it isn't going to be as uniform as you might think by looking at a sample. The uniformity of the oxidized surface is affected by the runoff of water, and the color will be different in those areas that are interior to the structure or that are shielded from the runoff. Rust staining is another issue if you're using concrete supports for the structural elements.

TOMASETTI: Do you think that the staining is one reason why the industry stopped using weathering steel?

FASULLO: There is a more significant problem that involves the technical understanding of weathering steel. First, the industry presented the quality of the final product based on a small sample of steel that showed uniform corrosion. We all know now that because of the runoff differences, the corrosion rate and the color are different; you don't get exact uniformity. Second, I think the issue of continued corrosion where the material remains wet has created a tremendous number of problems around the country. Many are so serious that they require either replacement or additional protection of the steel. While brochures may have had fine print advising of differences in corrosion rate due to environmental or atmospheric conditions, I really don't think the average architect could appreciate the impact of these words.

I think there is a place for weathering steel. For example, we successfully designed highways over the New Jersey Turnpike where the concrete is beneath the structure but all the runoff is directed away from it. If you compare those turnpike bridges with the earlier ones, the corrosion staining is minimal. Experience has taught us that you have to design to protect the concrete from the runoff or else accept the fact that it's going to be corroded and stained. I think you will see weathering steel being used in more judicious ways as a result of the last 30 years of experience.

TOMASETTI: What would you choose today as a parking structure?

FASULLO: As you know, there is no one answer to design; it all depends on the facts of the case, the competitive cost of materials, and the industry conditions that exist. We built a garage recently at John F. Kennedy International Airport, and after doing much analysis, we decided to use concrete. However, the architectural considerations were completely different than they were at LaGuardia. There is no one material that is ideal under all conditions. They all are merely options available to the architect and the engineer.

TOMASETTI: When my involvement in the industry began 20 or 25 years ago, I think there was more use of exposed concrete, especially for unique shapes, hyperbolic paraboloids, shells, and so forth. Why are we not doing as much of that now?

FASULLO: If we go back to, say, the 1960s, there was a much greater interest in the use of concrete shells because of their efficient use of materials. Also, shells have the beauty of really using a geometric form to carry the loads in a very direct way as membrane forces instead of as bending forces. The downside part of shells is the fact that wood forming is required to create them. So you're paying to build a fairly sophisticated forming system and then you're paying to remove it. The economics of formwork is what is making it less desirable today.

TOMASETTI: Every time I pick up a good trade publication such as *Engineering News-Record* and I see some really exciting structural system, it's in Germany, Australia, France, Spain. In America we make economics in our industry govern everything. Do you think we're seeing exposed structures in this country because we are getting so pragmatic relative to the Europeans and maybe even to the Asians?

FASULLO: I think that's particularly the case in the public sector. The public sector has an obligation to build structures not only that are functional and that meet the need of the particular facility but also that represent the value system of the society in which we live. When you build major airline terminals or bridges that may span the Hudson River for hundreds of years, one has to apply a higher standard. The real trick in the public sector is not only to be able to balance the two, but also to be able to communicate effectively and get support, because without both organizational and public support you won't get approval to build anything. On the one hand, you've got to be careful not to just be driven by the bottom line without considering the life-cycle costing, which sometimes requires a higher front-end investment but in the long term is economically justified. On the other hand, you have to be careful not to build something that may appear expeditious from an economic point of view but that may have no aesthetic virtues.

TUCHMAN: This gets into another issue of whether a good design costs more or whether a good design can be achieved regardless of the budget you use.

FASULLO: Obviously, no one would propose that a good design costs more in an absolute sense. And again, cost is measured in different ways. You have initial cost, operating cost, maintenance cost, financing cost, and life-cycle cost. Our first step in designing is to be really clear on our functional requirement. What do we want to accomplish? Then we try to assure that the cost is compatible with the criteria and function we are trying to achieve. It may

Figure 3.11 Exposed structure of bridge and bus terminal complement each other. **Port Authority Bus Terminal at George Washington Bridge, New York, N.Y.**; Architect and Engineer: *Pier Luigi Nervi.* (Photo: *Jerry Rosen, Port Authority of New York and New Jersey.*)

not be the lowest initial cost, but it has to be the lowest cost given the objectives of the project.

TUCHMAN: I think the word we are looking for here is quality.

FASULLO: Quality is a criterion. If you're building a marine terminal with a 20-year life expectancy, the definition of quality is different from one that applies to a monumental bridge or a tunnel under the Hudson River (Fig. 3.11). It's very important to define quality in measurable terms as part of the design process. When all the players participate, the final cost should be the appropriate cost.

TOMASETTI: As we talked earlier, there are two types of exposed structure. One is where the form of the structure is exposed but the material itself is not because it has been clad with architec-

Figure 3.12 Exposed steel trusses define this building while permitting open sides to exhaust fumes.
Port Authority Bus Terminal, New York, N.Y.; Architect and Engineer: *Port Authority of New York and New Jersey.* (Photo: *Jerry Rosen, Port Authority of New York and New Jersey.*)

tural-type materials. The other is where the structural material itself is exposed. Do you think it really works to make the material serve a dual function—to serve as structure and as architectural element?

FASULLO: I would say it is more economical to allow the structure to be the architectural expression, to serve a dual function (Fig. 3.12). The only premium you might pay, as we did at LaGuardia, is that the steel is detailed differently. In concrete, as we discussed earlier, you have a similar premium to pay on the basic structural form. If architectural finishes are not required, for example, to protect against the weather, to enclose the building, or to fireproof it, then I think those architectural materials are playing a separate and distinct role from the structure.

TOMASETTI: There are a lot of problems when you start exposing structure—temperature variations between the inside of the building and the outside, creep that sometimes causes movement, and so on. Do you think that over the past 30 years the advances in technology and in our knowledge to solve these problems is significant enough to encourage the use of more exposed structure in the future?

FASULLO: Those types of problems have always been with us; you have to anticipate, understand, calculate for, and compensate for them. I think one of the biggest shortcomings engineers have is that they think of structures in terms of stresses rather than in terms of strain. Strain is displacement caused by load applied. It doesn't matter what causes the strain; what is important is that we have a physical appreciation for the way structures move in their environment. Students should be taught more how to deal with strain rather than thinking just of stresses that come out of a computer. When the structure is exposed, the engineer has to be more physically appreciative and intuitive about the actual distortions, movements, and interrelationships.

Although the computer systems deal mostly with stress analysis, they actually can be used to display distortions and movements graphically. It's important in the educational process to focus equally on the distortion issue and on the stress issue so that overall building performance can be considered. Interaction of the total system has got to be focused on, and someone in the design team, in my opinion the structural engineer, must be the integrator of the overall behavior of the building.

HAL IYENGAR

Skidmore, Owings & Merrill has been responsible for a number of landmark buildings where the structure is exposed or expressed, particularly the 100-story John Hancock Center in Chicago. SOM partner Hal Iyengar, chief structural engineer in the firm's Chicago office, points out that SOM's unusual synergy in the development of architecture and structure provides good opportunities for expressed structures to emerge where appropriate.

Recently Iyengar developed the giant parabolic arches in exposed steel that carry a 10-story building across several sets of railroad tracks in London's Broadgate office complex.

TUCHMAN: Can you discuss the concept of exposing structure in terms of concrete and steel?

IYENGAR: Historically, fully exposed structure is more prevalent in concrete because it has qualities of durability, weatherability, and inherent fire protection. In the full exposure of structural steel, one has to deal with all the issues of fire protection and corrosion. You have to bring some of the issues of bridge technology related to corrosion protection into a building to expose steel.

However, the expression of structure—whether exposed or not—is something that has occurred quite frequently in both mediums. In these instances, the outer expression of the building reflects the presence and the character of the structure.

TUCHMAN: Like the John Hancock building in Chicago?

IYENGAR: Yes, the architecture fully reflects the structure and, in a way, the structure is the essence of architecture for the Hancock building. It is a very powerful system (Fig. 3.13).

TUCHMAN: So the structure comes first?

IYENGAR: In my opinion, yes. The architectural idea capitalizes on the fact that there is a structure. When you go back to the style of Mies van der Rohe, the tradition is to have as simple a structure as possible. So in some philosophies, like that of the International style and the Meisian style, the structure has a stronger presence than the architecture. When architecture is based on function and structure reflects that function, then you will have natural opportunities to expose or express the structure.

Current architectural styles have changed the situation in regard to structural expression. You have a certain amount of a priori attitude toward architecture—that it has to have rich exterior articulations and decorations. It becomes very difficult to provide for offset profiles in evolving a logical structure.

TUCHMAN: But going back, how did the Hancock building come to be?

IYENGAR: There the issue was very simple. We had a multiuse building with commercial space, parking, offices, and apartments. One idea was to build two buildings on the site, one for offices and one for apartments. The width of an apartment building tends to be narrower because you can't plan too much interior space in apartments.

TUCHMAN: How did you solve the problem?

IYENGAR: In this case Fazlur Khan [Skidmore's brilliant chief engineer who died in 1982] came up with the idea that we could cre-

Figure 3.13 Wind cross bracing is fully expressed as an integral part of the visual design of "Big John."
John Hancock Center, Chicago, Ill.; Architect and Engineer: *Skidmore, Owings & Merrill.* (Photo: *Hedrich-Blessing, courtesy of Chicago Historical Society.*)

ate one structure, and that the structure could be based on a continuous diagonalized tube. Only then did it occur to the architect that it was possible to have a very strong structuralist expression. The structure comes first to see if there is a natural efficient structure. Once the natural structure is there, then we consider whether it merits a strong role in the architecture.

TUCHMAN: It sounds like a fairly organic process.

IYENGAR: Exactly. There are very few buildings that really start out when someone says, "Oh, yes, we will create a structuralist expression. Let's go at it!" I don't think it happens that way. It happens because the characteristics of the project create a unique structure and then the architect capitalizes on it. That's exactly what happened in the Hancock building.

TUCHMAN: How significant was Faz Khan's contribution?

IYENGAR: His work was enormously helpful in understanding structuralist expression, because Faz always came up with very efficient structural systems. Obviously, they had to evolve for structural reasons first, then they had the ability to set a structuralist context for architecture. He was bold enough to come up with exterior structural forms and challenge the architect to exploit them in the architecture.

He came up with a series of ideas while working with students at the Illinois Institute of Technology in Chicago. He'd have these ideas in the back of his mind, and when SOM got a project, he would propose one and see if there was an architectural reason for it. Through this system, he was able to put forward structures that immediately had an impact on architecture.

TUCHMAN: Can you give me an example?

IYENGAR: Faz came up with several ideas for the tubular building, and then a lot of buildings were built with the closely spaced columns of the tubular vocabulary. Of course, the classic tube is the diagonalized tube, and the opportunity to do that was the Hancock building. The diagonals go all the way around and form the steel tube, and the building tapers at the top because the apartments need to be smaller.

TUCHMAN: At SOM, of course, you have the added dimension that the architect and the engineer are partners in the same firm.

IYENGAR: Yes, a great deal depends on the chemistry between the engineer and the architect. One has to be able to share from the first day that we have this building, this kind of program. What are the structural possibilities? What are the architectural possibilities? Then you develop things.

At SOM we are literally in close proximity to the architects in the studios. There is nothing that goes on from the first line that's drawn we wouldn't know about. We put forward ideas very quickly before the architecture has taken shape, and then they have the opportunity to become an integral part of the solution.

TUCHMAN: At Onterie Center you expanded the vocabulary of the diagonally braced tube to include concrete (Fig. 3.14).

IYENGAR: Yes, and at Onterie Center the reinforced-concrete diagonally braced tube is expressed and fully exposed. In architectural concrete, the concrete itself is completely exposed to the elements. But we are coming to the view that we need to weatherproof by painting exposed concrete structures because of durability problems—something we never thought about 20 years ago. We did a couple of unpainted ones, and we've had to go back with inspections and maintenance. Unless some kind of sealer was applied on the outside periodically, there would be corrosive effects. Concrete will crack, and moisture will penetrate and freeze and work its way into the reinforcement, which will rust. Every few years you have to go back and resurface. Then you try to patch and have color problems and all those kinds of things.

TUCHMAN: How is Onterie Center holding up?

IYENGAR: Onterie Center is holding up well. But it is painted. Generally there won't be any effects for 15 or 20 years, but then you'll start to see it. Though, actually, the Gateway Building, built in Chicago in the early 1960s, doesn't show any deterioration yet. It depends somewhat on the detailing of the reinforcement and whether or not there's enough cover for the reinforcement.

TUCHMAN: Your most recent exposed structure is at the Broadgate Complex in London, the building carried on parabolic arches across the railroad tracks (Fig. 3.15). How did that come to be exposed?

IYENGAR: It's a very daring building—direct structuralist expression with exposed steel. The owner, British Rail, originally thought the site was unbuildable. The concept was generated by the placement of a building on top of the railroad tracks. We couldn't use too many columns, so we had to span across many tracks. We created an idea for a structure that spans 250 feet and carries 10 stories. Another attraction here was that the building faced a historic railway station, a beautiful building, a gateway that recalls the grandeur of the old railroad era.

It made sense to pay homage to the structuralist gateway, but create a building with merit of its own. In this design, the struc-

Figure 3.14 By filling selected window openings, this concrete diagonally braced frame has structure expressed and exposed.
Onterie Center, Chicago, Ill.;
Architect and Engineer:
Skidmore, Owings & Merrill.
(Photo: *Greg Murphey.*)

Figure 3.15 The tied arches carrying this building over railroad tracks are not only expressed and exposed, they also stand free of the facade. Fire protection received particular study.
Broadgate Complex, London, England; Architect and Engineer: *Skidmore, Owings & Merrill.* (Photo: *Hedrich-Blessing.*)

ture is very clearly expressed; it is a philosophically pure, highly ordered structure. There is not one item that's extraneous, or that's been distorted in any way from the structure. But, of course, we had to come up with a coating and paint system that maintains the durability of the steel, and we had to go through a tremendous amount of fire engineering studies to make sure we met the fire protection requirements (see Chap. 5).

There was also a great deal of work relating to the structural engineering, and a great deal of consideration given to the member proportions and aesthetics.

TUCHMAN: Talk about the structure a bit more. How does it work?

IYENGAR: The building has three spans: 18 ½ meters (60 ½ feet), 15 meters (49 feet), and 18 ½ meters again. And there are four arches: two on the interior and two actually outside the building to create these spans. These arches suspend a series of hanger columns to which the floor system is attached. Some are hangers below the arch and become columns above the arch.

TUCHMAN: Do these columns rest on the floor or carry the floor's transfer?

IYENGAR: The floors transfer load to the columns and the columns transfer load to the arch, and then into the abutment.

TUCHMAN: Two arches are exposed and two are not?

IYENGAR: Yes, and for the outside arches, an open space is provided between the structure and the glass line. It serves two purposes. First, it projects the structural arched frame on the outside, so you can see the structure very clearly. Second, because the structure extends outside the curtain wall by 2 meters, the temperature effects resulting from the fire can be kept under control. This is the essence of new fire engineering technology that is emerging.

However, we did use fire-resistant glass since we could not fully satisfy temperature requirements. We thought we had to come up with a series of flame shields, like it was done in the One Liberty Plaza building in New York (Fig. 3.24), to control temperatures.

TUCHMAN: But you didn't use them?

IYENGAR: No. Our initial design included a lot of flame shields, but in the end we decided not to use them, because there were questions about their durability as well as about the aesthetics involved. This is the kind of building where you see all the bolts, all the pins, all the details of the structure.

It's literally designed as a bridge incorporating 10 stories. We have support abutments with sliding bearings on one side and fixed on the other. The arch system involves the arch, the hangers, and major and minor ties.

TUCHMAN: Would you say a few more words on interaction? Where does the owner or developer come into the process? What kind of reaction do you get when you talk about exposing a structure?

IYENGAR: In this project, for example, the rest of the Broadgate complex of buildings has a very heavy masonry facade, to be contextual with its London neighborhood. In the center of this complex we have a very special building with a jewellike character and an open plaza. This concept was immediately embraced by the owners because they were looking for some kind of structuralist centerpiece.

But reaction varies from client to client. When we did the Hancock building in Chicago, there was a great deal of concern about where these diagonals would fall and whether these spaces would be marketable. It was thought that the diagonal going through the space could be a detraction. But the fact that we could create a 100-story building for about the same cost as two 50-story buildings is what swayed them to go ahead with it. As it turned out, the diagonals are not liabilities, but assets, especially in the apartments.

TUCHMAN: That's what I would say, as well as in terms of the building's image, its whole character.

IYENGAR: It's good.

TUCHMAN: You say even the apartments?

IYENGAR: Yes, people have incorporated the diagonals into the designs of their apartments. Some use them for inclined bookshelves, some arrange paintings around them—in general, they have a sculptural effect. So they are definitely an asset.

But there are clients who, particularly now, view this type of system as too much structural order, too much of a statement. Having built the Hancock building, we may not be able to do the same thing again very well. Another client will say about a design, "This looks too much like the Hancock. We want our own image." So it becomes difficult to use that kind of system with that exact vocabulary, but there are other ways to do it.

TUCHMAN: You mentioned context in regard to Broadgate. What do we build next to the Hancock building?

IYENGAR: Something different, I'm sure. It just so happens that Hancock is unique. It's sitting in an area that's residential as well as commercial, so there are not too many office buildings nearby.

TUCHMAN: I'd like to touch on cost for a structure that's exposed versus one that's not, or, say, one that's expressed. Do you feel the savings outweigh some of the added costs? For example, with Broadgate, you had to take special precautions on fire protection; at the same time, you didn't have as much cladding.

IYENGAR: In general, I think it has turned out to be more economical to expose the structure. To start with, you want a more efficient structure, so there are economies there.

TUCHMAN: Does the expense of designing an exposed building become greater because it takes more design hours to figure out the details of exposing it?

IYENGAR: Yes and no. We've certainly spent as much time designing very traditional buildings. If you are doing anything special technologically, you need to do the investigations. You need to be sure that when you bring your design to the client, he won't ask you for a lot of backup to make certain the design has no structural flaws. So there's some additional work required, though in most cases, I don't feel it's a big factor.

TUCHMAN: But overall, your sense is that it is more likely to be an economic method?

IYENGAR: It also depends on the scale of the structure. Sometimes it takes a great deal of work to make a very small building a structural jewel, but the fee is small because the scope of the project is small. There it would be somewhat out of proportion. But if the scale of the project is large, I think you can organize things so that the fee is a reasonable return for the amount of work you put in.

TUCHMAN: You seem to agree that the term "exposed" means that the surface has no insulation or cladding, but that it may be coated. That brings up the question of 780 Third Avenue in New York.

IYENGAR: That building, now called the Wang Building, uses a system of diagonal concrete panels for bracing, similar to the Onterie Center (Fig. 1.2). But the New York building is not exposed. Because of the building's image, they added granite cladding to make it look a little more sophisticated. And because the window system is so similar in color to cladding, you can hardly see the Xs. The two buildings have very similar structural systems, but Onterie is exposed and Wang is not.

WILLIAM LEMESSURIER

A 1989 symposium in New York City called "Bridging the Gap: Examining the Relationship between the Architect and Engineer" set the stage for an interview with William LeMessurier, chairman of William LeMessurier Associates, Cambridge, Massachusetts, and an adjunct professor at the Harvard Graduate School of Design. LeMessurier talks about dramatic cantilevers he has designed at the Citicorp headquarters in New York City and for the Singapore Treasury Building, where circular floor plates are individually cantilevered from the core. LeMessurier was also the structural engineer for the Boston City Hall. He calls it a "celebration of structure" because of its extensive use of exposed concrete.

THORNTON: We think, from our own experience, that to truly expose or express a structure costs more during the design phase of the project. Not only is it more labor intensive for the architect and the engineer, but also the engineer has to take the lead to prove to Mr. Owner and Mr. Architect, in certain cases such as Citicorp, that it does work—it does fit the module. There are a lot of times I find that it is not financially advantageous to expose the structure. What do you think?

LEMESSURIER: You are obviously correct, but if I ran my whole life just to maximize the profits, I wouldn't have accomplished anything.

THORNTON: We've done that too. In the 1950s and 1960s it happened, when we were one of the primary long-span structural engineers in the country. But we got away from it because we couldn't make any money. With everybody else's fees, the structural engineer isn't left with enough to do the job properly and also make money. One reason for this is that on long spans there's more engineering; the other reason is that the architecture and the structure become one, and you end up spending more on the project. I agree with you—it gets on the cover of magazines, it enhances your personal reputation, and you feel good about it. But it does take more time to do a well-done exposed structure.

LEMESSURIER: The older I get, the wiser I get, and the more I think I might agree with you. In the end, however, I design for myself, which means for my own intellectual satisfaction, and I don't really care whether anyone sees it or not.

THORNTON: After the lecture last night, several engineers were talking about Citicorp. One of the beauties of the Citicorp headquarters in New York City is that you look at the building and you see the column in the middle and the horizontal, but not how it's truly supported. It's indirect; it's subtle—you can't tell. If you and Hugh Stubbins had put big diagonals on the outside, it might be stating something that Stubbins didn't want to state. Maybe he wanted to leave the perceiver in a little bit of confusion, because you can't tell from the outside, unless you look very closely, that you have the chevron diagonals (Fig. 3.16).

LEMESSURIER: Stubbins really was adamant about keeping the structure inside the skin, and not because he didn't like it or didn't want to see it. It would have cost an arm and a leg to see it from the outside. The only building that really has the structure outside belongs to an aluminum company, because they had the aluminum sheathing to go around the entire structure. That's spending money for art. If you've got that kind of money, then go ahead and do it. But if you're going to build a no-nonsense developer's building, which is what Citicorp was supposed to be, then you cannot. The only reason Citicorp has all the drama of being lifted up in the air is because that was the condition of buying the land. And there's no question, as Kevin Roche would say, that it's a structural spectacle (Fig. 3.17).

THORNTON: So, in other words, you're concurring that exposing the structure and then having to cover it is more expensive than putting it back inside.

LEMESSURIER: But we're talking about steel frameworks in big

Figure 3.16 This structural elevation, typical for four faces, shows diagonals collecting vertical load to a mid-face leg. This pattern is visible only when backlit at night.
Citicorp Center, New York, N.Y.; Architects: *The Stubbins Associates with Emery Roth & Sons*; Engineer: *LeMessurier Associates*. (Drawing: *The Stubbins Associaies.*)

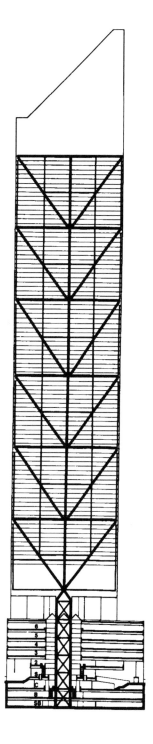

Figure 3.17 Four legs and a central core lift the building to clear space for the church at left, but still give a feeling of stability. Cross-street reflections create the light variation on the skin.
Citicorp Center, New York, N.Y.; Architects: *The Stubbins Associates with Emery Roth & Sons*; Engineer: *LeMessurier Associates*. (Photo: *Edward Jacoby*.)

office buildings. I think there's another whole world out there. I have to say that throughout the 1950s and for a good deal of the 1960s, all the work we did was exposed structure in one way or another. And a large part of it was reinforced concrete. There was a passion for doing exposed concrete structures. The biggest example is the Dallas/Ft. Worth airport. Everything is there. It's made of great big chunks of beautiful brown concrete that has been sandblasted; posttensioned components are the main frame of the building. In between these, I actually made the floors out of steel on the grounds that the airport would be constantly remodeling and it would be possible to cut new holes and put in, for example, new baggage systems. That's the way it worked, and it worked quite well.

THORNTON: I think what you're saying is that we've gotten away from those kinds of things because the industry doesn't want to do them. It's very hard to get an exposed concrete job to fall within budget, right? The tendency now is to put up a structure and cover it.

LEMESSURIER: Yes, but it's not difficult to get precast. As you know, we build with it all the time, and some of the constructions are very nice. The architects themselves have gone off onto this tangent of historicism and decorativism. Why don't we stop slathering this granite all over everything? I mean it costs a lot of money and doesn't look too good; everybody has to have a granite skin on their building. Why don't we go back and put on bricks? After all, the Chrysler Building was built with brick. Go look at it if you've forgotten. Actually, you can put moldings and things in there, and it's beautifully crafted.

THORNTON: So the postmodern movement in a way has sort of steered us away from the fashion of exposing structure...

LEMESSURIER: No, no, no, no. The pendulum is beginning to swing the other way. We should touch on the very important question of temperature control in an exposed structure. There's a very simple solution to it: don't put any structure inside the envelope, and make sure it's all out there and it's all getting cold or it's all getting hot. There were a lot of people that didn't understand that at the beginning.

THORNTON: We did a six-story nursing home once in New Rochelle with an exposed concrete column, and we didn't think about it. The architect went ahead and put in partitions separating rooms from the inside to the outside, and every one of them had a diagonal tension crack in the drywall partition.

LEMESSURIER: We have no problems with our own little six-story building, with its painted concrete. That's a terrific material. It really works. When we put the paint on, the guy that worked for Stubbins was inspecting the paint. He said, "What kind of paint do you put on concrete?" "The same paint they put on all the stone at the Empire State Building," was the reply. Go and look at it. It's not limestone; it's all paint.

THORNTON: Your building is the one on Massachusetts Avenue (Fig. 3.18). I know it's concrete, but I never realized it was painted concrete. How many times has it been painted since it was constructed?

LEMESSURIER: Only once, and it's now over 25 years old. It's a good example of a no-nonsense, practical, exposed structure. And people don't even realize it when they see it. It's very simple, very cheap.

TUCHMAN: Let's go to a much more complex structure, the Singapore Treasury Building. Would you call that an exposed structure?

LEMESSURIER: Not really. It's all covered with granite or something at the bottom. There is a part of it that shows, but at the lowest level of the building half of it comes down to the ground and the other half is cut off. At the arrival point, there is a great glass space frame roof that's out in front. You can see that wall, but that's about it; I don't think you can see the concrete (Fig. 3.19).

TUCHMAN: So you can't see the cantilever either? It's not expressed?

LEMESSURIER: Oh, the cantilever is over your head, just like the Citicorp one. When you're inside, the mullions around the perimeter are exceedingly skinny, so you get this breathtaking...

THORNTON: Is each floor individually cantilevered?

LEMESSURIER: Individually.

THORNTON: But there's hung ceiling throughout.

LEMESSURIER: Yes and no. The ribs are expressed, and the ceiling rises between them. The form of the structure is very clear in the reflected ceiling plan of every floor.

THORNTON: From the outside you can look at the building and see the thin core rise and disappear into the tower, so at first a person would obviously conclude that it's all taken into the core. And when you walk into the building and stand on a typical floor and look out, you see in the ceiling the gradually thinning...

LEMESSURIER: No, they're absolutely flat because they wanted the ability to cut off spaces, although I don't think they've done very much about that. There really wasn't much point in thinning it.

Figure 3.18 With good workmanship, exposed painted cast-in-place concrete framing like this can give the appearance of limestone. **1033 Massachusetts Avenue, Cambridge, Mass.**; Architect: *The Stubbins Associates*; Engineer: *LeMessurier Associates.* (Photo: *The Stubbins Associates.*)

Figure 3.19 Half this building is interrupted above the lobby, as the upper floors are cantilevered out.
Singapore Treasury Building, Republic of Singapore; Architects: *The Stubbins Associates and Architects 61*, Singapore; Engineer: *LeMessurier Associates and Ove Arup & Partners*. (Photo: Courtesy of *The Stubbins Associates*.)

We have a fair number of duct holes and it's easier to make a hole through a deep member, as you know.

TUCHMAN: Is there some spandrel panel, or is it floor-to-ceiling glass?

LEMESSURIER: It varies. The architect made the windows on one side of the building higher and smaller than on the other side. This was really so that you'd have some sense of directionality when you were in the building. When you park around the building, you know whether you are on the north side or the south side, but you can't tell that from the sun because the sun is straight overhead since the building is so close to the Equator. You also gain a certain richness on the outside.

THORNTON: But there was a spandrel panel.

LEMESSURIER: There's no structure above the floor. There is the panel, aluminum insulated.

THORNTON: Because of the radial layout though, the span of the tip gets quite large and you use the space truss.

LEMESSURIER: We had a truss that was designed for a little redundancy. If the moment connection ever were to fail on the inner end of any one of those 16 trusses, it still could hold its shear capacity and the entire gravity portion of half of that thing could be transferred to other elements.

THORNTON: Very smart.

LEMESSURIER: It also stiffened the edge for differential live loads on the floor so that the floor wouldn't be disturbed very much at any particular point. We felt that was quite important.

THORNTON: Did you also use the structural mullion to take out the vertical differential?

LEMESSURIER: We never intended to, but we were ordered to do so by a man who was on the board of directors at the treasury and who was chairman of the civil engineering department at Singapore University. He said, "You've got to do something," so we developed the thin flat plates that interconnected the floors. If something went down, it would never put any load on the plate below it because there was a sliding joint so it couldn't buckle. It could only deliver load up—just floor by floor, not all the way to the top.

We also have the most reliable moment connection the world has ever seen. I didn't talk about all of this last night, but the connection of that steel to the core was accomplished by having a vertical built-up column that was erected first; the concrete was poured later. It had a moment connection made in the shop with a bracket that stuck out about 6 feet and was spliced between the main cantilever. That was made in the field with bolted connection plates on the bottom. Because they are subject to lamellar tearing, the special plates required welds to get through to the stiffeners. To prove the quality of the steel, tiny $\frac{1}{16}$-inch holes were drilled through it at roughly 4-centimeter intervals and then examined under a microscope.

We just put that steel in the region where this connection was made and went back to ordinary 850-pound steel for the rest of it. In addition, there were braces, which were ultimately buried in the concrete, that were made of angles to stabilize the whole thing. Then the concrete floor was cast, and the moment, the dead-load moment, was carried into the steel and resolved there. Then the concrete wall came up so that the final restraint is meter-thick concrete, which has enormous rigidity to it.

TUCHMAN: What building have you worked on that exposes the structure more extensively?

LeMessurier: The Boston City Hall is the most thorough and extreme example of exposed structure that I've every been involved with. Every foot of it is outside surface. It's a celebration of structure (Fig. 3.20). Although I helped to figure out how to make it happen, the architects (Kallmann, McKinnell & Wood; Campbell, Aldrich & Nulty) conceived it all. There's raw concrete poured in place, and the main columns were done with a beautiful board finish. The boards were used only once. It was slow in building, I'll tell you. I really don't recommend it as a model for behavior, but, on the other hand, it's an extravagant piece of exposed structure.

THORNTON: And what's nice about it is that it's got a very nice combination of precast and cast-in-place concrete. The outside that you see is mostly cast in place, isn't it?

LeMessurier: Oh, the upper part of the building is all precast.

THORNTON: Okay, but there are some really big, thin walls...

LeMessurier: Those thin walls are cast in place. When you get up in the building, the wall starts moving out, cantilevering in a sort of corbel, and that corbeling is all done with precast pieces that contain mechanical systems that go up and get larger near the top. But all of the pieces out there are structural, precast working pieces as well as the finished skin. And it has stood up quite well. It's 25 years old now (Fig. 3.21).

TUCHMAN: The other building I want to ask you about is Gyo Obata's spiral concrete shell (Chap. 2). I think what is intriguing about it is that the building looks as if it ought to be an exposed structure, a fluid concrete shell, and yet there was at least talk in

Figure 3.20 This "celebration of structure" combines exposed cast-in-place concrete walls with precast mullions.
Boston City Hall and Plaza, Boston, Mass.; Architects and Engineers: *Kallmann, McKinnell & Knowles; Campbell, Aldrich & Nulty; and LeMessurier Associates, JV.* (Photo: *Len Joseph.*)

Figure 3.21 A detail shows cast-in-place work (with carefully spaced tie holes) mixed with precast soffits and mullions.
Boston City Hall and Plaza, Boston, Mass.; Architects and Engineers: *Kallmann, McKinnell & Knowles; Campbell, Aldrich & Nulty; and LeMessurier Associates, JV.* (Photo: *Len Joseph.*)

the beginning that it was going to be a steel frame.

LEMESSURIER: No one could afford to build it out of concrete. Our role in that was very specific. We were paid a fee for the conceptual design of the structure. I understand they were going to do the details in-house.

THORNTON: What is it, a light steel frame covered with plaster, stucco, or...?

LEMESSURIER: It will be covered with sheetrock. The form of the building itself is so dramatic that getting too excited about the detail of any particular beam is beside the point. I mean it's a building that's a classic logarithmic spiral. It starts at the top, and it gets bigger and bigger and finally ends up in this great sweeping circle at the bottom. It's like a dramatic elongated sea shell. One of the problems with it is that the curvature is constantly transient; there is no such thing as radius of curvature. If you try to build that thing out of concrete, you're going to build the most monumental formwork bowl that can only be used once. You can't take a panel here and move it there.

THORNTON: It's like the Eastman Kodak Pavilion at the World's Fair. There was no mathematical formula that defined the curve. We generated it from a model.

LEMESSURIER: Actually, the curvature in this thing is quite rigorously mathematical.

TUCHMAN: Has an architect ever come to you looking for drama in exposed structure?

LEMESSURIER: It happens all the time. I recently had just such an experience. One of my favorite people is Ben Thompson, whom I have been working with for 30 years. He got a commission to redo the central town square in Rotterdam, which has an opera house on one side, theater on the other, major shopping, and parking garages underneath. It was awfully deadening—there wasn't a tree or bench or anything. Anyway, they decided they needed a covered walk to go diagonally across the square, and they asked for our ideas on it. They really intended just to keep the rain off people's heads. So we suggested putting posts at 10- or 20-ft centers. They didn't want to do that. "We want something dramatic." We ended up with a kind of lemon-shaped truss that goes across three-dimensionally like a big Tootsie Roll. It's squashed, covers the space, and sits on some tripods which then have masts going up. We put wires on to help it out because this thing is 400-odd feet long and spans about 250 feet. Anyway, we built a model of it, and they absolutely fell in love with it. They had a view of it that I hadn't had in my mind at all. They said,

"This is the spirit of Rotterdam. That is the most characteristic scene. It's the greatest harbor in Europe, with all these ships going up and down." It was sold, just like that.

I don't think owners usually come to us for all these aesthetic things, however. They come to us because we have a reputation for things not falling down and for doing economical things. It's a combination: reliability and no-nonsense economy.

MATTHYS LEVY

When Matthys Levy moderated an AIA panel discussion in May, 1988, he went to the heart of the issue of exposed structure. He believes exposed designs celebrate the romance of structure—not just the cold logic of geometry and the strength of material, but something that goes beyond and contributes to the architecture medium. As President of New York City-based structural engineer Weidlinger Associates, Levy has designed a number of exposed structures, influenced by his now retired partner Mario Salvadori—author of a well-known book on structure as architecture. Particularly noteworthy is the progression of exposed designs, from Saarinen's spandrel beams to I. M. Pei & Partners' spectacular space frame envelope of the Jacob K. Javits Convention Center, and now to a stacked-ring system of cables and posts for the Georgia Dome in Atlanta (Figs. 3.22 and 3.23).

LEVY: Let's start with the question, "What is an exposed structure?" I think it's any structure which is not covered up—it's as simple as that. It doesn't make any difference if it's inside or outside—and I don't consider a coat of paint the same thing as a coat of cement. Take a steel structure like One Liberty Plaza, where the steel girders are painted. They are exposed on the outside, so that's an exposed structure. The columns aren't exposed because the fireproofing requirements wouldn't permit us to do that. Visually, the impression was that the two are identical (Fig. 3.24).

TOMASETTI: On One Liberty Plaza in lower Manhattan, where you have a direct flame shield around each plate, you go further and say that even if you put something like that on the structure, as long as you maintain that form of structure, you consider it exposed.

LEVY: No, it's an exposed structure because the web is totally exposed. The web is the structure, the flange is covered up by fireproofing and another steel cover on top of that.

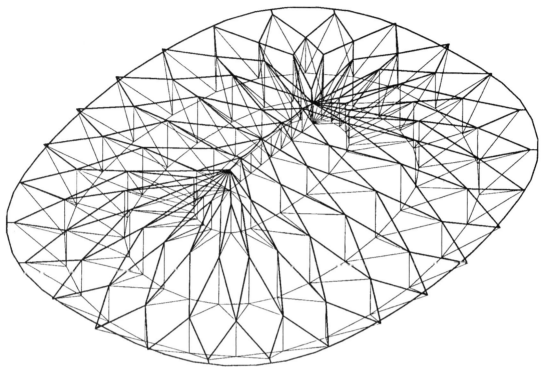

Figure 3.22 The stacked-ring and three-dimensional nature of framing for this dome is clearly visible. Heavy lines show the perimeter compression ring and the diamond-patterned layout of upper tension tendons. Light lines indicate vertical compression struts and the sloped tendons used to lift the fabric roof.
Georgia Dome, Atlanta, Ga.; Architects: *Heery/RFI/TVS Joint Venture*; Engineers: *Weidlinger Associates and Harrington Engineers*. (Drawing: *Weidlinger Associates*.)

TOMASETTI: That was a flame shield.

LEVY: Yes, I think there are more exposed concrete structures around than steel, simply because of fireproofing problems. We've done tons of them around the country, in Atlanta, Dallas, Des Moines, all totally exposed. What you see is what you get...no other architect is involved...the form of the structure is the form of the building, and the detail as well.

There are two kinds of exposed structure. You can have an exposed facade in tandem with an exposed interior, where, for instance, the floor system is exposed. Twenty years ago we did some buildings with Skidmore, Owings & Merrill that were all very similar to each other, with exposed precast concrete floor systems. One of them, American Can in Greenwich, Connecticut, has precast tees and beautiful round ducts that go in the space

Figure 3.23 Fabric is being installed over the inner stacked ring. The tension ring with catwalk, the vertical posts, and the diamond-patterned cable layout are visible.
Georgia Dome, Atlanta, Ga.; Architects: *Heery/RFI/TVS Joint Venture*; Engineers: *Weidlinger Associates and Harrington Engineers*. (Photo: *Richard Gore.*)

Figure 3.24 These steel spandrel girders are exposed but required flame shields at flanges.
One Liberty Plaza, New York, N.Y.; Architect: *Eero Saarinen*; Engineer: *Weidlinger Associates*. (Photo: *Bo Parker*.)

between the tees, with the light above that, so the whole ceiling is a beautiful integration of structure in exposed mechanical systems. This series of buildings used the same kind of structural, mechanical, architectural expression, all designed by Gordon Bunshaft.

Then there is another family of buildings, where you deal only with the skin. The first was in Brussels, and it has beautiful lacy tee-shape members with stainless-steel connections. That same type of expression was repeated in a number of other buildings in London, Houston, Kansas City, New Orleans.

TUCHMAN: What about the Jacob Javits Convention Center?

LEVY: That's a special structure, in a completely different category from the concrete-based buildings I've dealt with.

TOMASETTI: It's definitely exposed structure...a total form giver.

LEVY: It's totally there, but it's different in the sense that it doesn't have to satisfy a commercial function. As an exhibit space, it's much easier to develop as exposed structure. It's essentially a one-story building with extremely high ceilings (Fig. 3.25). It has dramatic impact, but it's different from the commercial and office buildings I've worked on, where the structure was based at least partly on economy.

TUCHMAN: This brings up a key question: does the structure become less expensive because you are leaving out material, or more expensive because a visible structure means everything has to be detailed to the nth degree?

LEVY: It's true that a lot of exposed structures cost more, but it's the structure that becomes more expensive, not necessarily the overall building—which can still end up costing less.

TOMASETTI: Would you say it depends on the way you use exposed structure? In other words, if the structure is somewhat more expensive, but it deals with function very successfully, you may end up saving somewhere else—say the architectural finish.

LEVY: That's right. For example, the double-tee type of exposed ceiling costs a little more—not even that much more—but because you use a deeper structural element, which is integrated with the mechanical duct in a very clever way, you could use a very cheap light fixture. It was just an exposed light bulb above the duct, so you didn't see it and the structure became a light fixture itself. You save on those diabolically expensive fixtures, and there was no hung ceiling, so over all it was less expensive.

TUCHMAN: Now this was done some years ago. Have you seen it again?

LEVY: No. For one thing, it works very well when you have large open spaces, like bank buildings and insurance companies. It

Figure 3.25 This three-dimensional space frame wraps from walls to ceiling and uses steps to advantage as stiffening elements.
Jacob K. Javits Convention Center, New York, N.Y.; Architect: *Pei Cobb & Freed Partners;* Engineer: *Weidlinger Associates.* (Photo: *Weidlinger Associates.*)

doesn't work so well with a divided space, so from that point of view, it's not very flexible in terms of architectural demands. But also, styles change. Exposed structure currently suffers from an image problem. After coming out of the closet during the modernist era, it has been slowly pushed out of sight by the postmodernists. However, some types of buildings, namely high-rise buildings and long-span structures, do not so easily permit this regressive attitude. They allow structure to participate as a form giver.

TUCHMAN: Are you doing any projects now with exposed structure?

LEVY: It's not as prevalent today as it was—postmodernism really does not lend itself to structural expression. Postmodernism is a facade, and the facade is not necessarily the structure itself, it stands in front of the real thing.

TOMASETTI: What would you say is the driving force in exposing a structure? Does it come out of some interaction between the architect and the engineer or does the architect wake up one morning and say, "I think I'll work in exposed concrete?"

LEVY: In my own experience, every exposed structure has been due to tremendous interaction between architect and engineer. The convention center, with the interaction between Jim Freed (Chap. 2) and myself, is a very dramatic example of that. It began as a sort of give-and-take over the original plan organization and what the structure should accomplish. We wanted a universal type of structure that would work in the flat and would work on vertical surfaces. To accomplish that, the geometry had to become the dictator, and we worked at a number of geometries until we could settle on one that seemed to satisfy all demands.

You had to be able to change the layering of the structure. In certain areas it would become deeper because its forces are greater, and you want to express that because you need the extra depth to carry the loads. The final geometry came out of the combination of aesthetic and structural demands. For the space frame, we studied a number of different frames and decided on one with nodes that were round balls and bars that were tubes.

So we said, "That's what we want: we'll do a structure with those two characteristics," and there were a number of patented types of structures that could accomplish that. We constructed a set of specifications that made the manufacturer responsible for responding to certain rules we had set up—sizes of members, the type of steel we wanted, and how it was to be used.

TUCHMAN: How much does the owner have to do with the initial concept?

LEVY: I think it's rarely the owner's original idea, but there are some sophisticated owners around with very definite notions on what they would like to see, and who often have strong input into the design. But in the case of the convention center, it was the architect who wanted to create a modern version of the Crystal Palace.

TUCHMAN: When an architect has an idea and the engineer says "we could expose that structure," does the owner ever say, "I don't like it?"

LEVY: Very often—but I've found if you make an economic argument, that settles the issue. "So, yes, I'm giving you an exposed structure, but do you realize it will cost you less than if I did it another way?" He says, "Okay, I'll buy it," and that's what it really comes to. Architecture is salesmanship. You can't win aesthetic arguments. If the guy says, "I don't like it," you're not going to win.

TOMASETTI: What about the durability? I've seen exposed concrete that doesn't look so good 15 years later.

LEVY: When architects first started using exposed concrete in the 1940s and 1950s (the loveliest of all were the tall buildings done in Chicago and elsewhere), the engineers found they were not aware of some of the problems, mainly thermal problems. A lot of the work that Mark Fintel did at the Portland Cement Association during that time was directed toward resolving these issues. Engineers had to be made aware that you're going to get differential temperature expansion, not only between inside and outside columns but within the outside columns themselves. Because one face of the column is hot and one face is cold, a gradient occurs within the column itself. So you had a global as well as a detail problem, and both resulted in some failures, in terms of cracking and peeling. We're discovering today that parking garages carry all sorts of problems (Chap. 7). We're encountering stupendous problems with chloride in garages that are less than 10 years old. That's an instance where the owners don't want to put a protective coating on the floor for, say, a dollar per square foot, and 10 years later the place is falling apart and he's spending many, many times the cost of the coating to repair his floor.

TOMASETTI: What about the engineer's responsibility as far as identifying things the owner should consider that may be above the minimum?

LEVY: I think you have a very clear responsibility if you have a potential problem that could be alleviated either by maintenance or by the addition of some other protective material. There, it's chiefly a question of information you should give the owner. If it's an issue of safety, then you have a responsibility to raise more than a red flag.

TOMASETTI: What's your current thinking about the use of weathering steel as exposed structure?

LEVY: We have one large building, a medical center in Brooklyn, which was done in exposed weathering steel, and it's a disaster. Fortunately it was not an exposed structure as such—the weathering steel was a curtain wall, a facade. The curtain wall has some serious problems in terms of degradation; joints don't work; sliding joints or connections are all locked in. There wasn't enough understanding of the rusting process to identify areas that would not rust in the proper way.

TOMASETTI: Do you see much future use of weathering steel in exposed structure?

LEVY: I don't think so. I think we'll see a lot more painting going on. Just like they painted the building at LaGuardia Airport—the whole parking structure there is now painted.

TUCHMAN: Is weathering steel an example of a good idea that didn't make it? Does the material just not have the full properties it was believed to have?

LEVY: No, I think it's a material that does exactly what it's supposed to do: rust in a controlled fashion, under certain conditions. It's the conditions that were not completely understood or properly applied. All moisture has to be able to evaporate, so the weathering action occurs.

You can see it clearly when you look at a lot of the bridges. There were a number done in New Jersey with weathering steel. Look at the difference between the lower part of the web, which rarely gets a lot of rain and direct sun and drying and humidity, and the upper part, where you see a slightly different color. They do not get the same weathering action.

TUCHMAN: On another topic—do you think it takes more design hours, and therefore a higher fee, to produce a successful exposed structure?

LEVY: No question, because you have to spend a lot more time detailing the structure. You're more involved in developing details that both satisfy the structural requirements and have some aesthetic appeal. You can detail very coarsely and it will work fine, it just won't look very good. On the convention center, especially on the curtain wall, there were a lot of choices involved. You could choose a rectangular gusset plate, a square gusset plate, and then we ended up with semicircular gusset plates, simply because they look good and they offered the lack of directionality we wanted to achieve. But it takes time to really work those things out.

TOMASETTI: Do the owner and architect appreciate that when you give them your proposal?

LEVY: No. The architect appreciates the time you're spending on it, but he's not going to pay for it. You can only convince them that it's worth the effort when either you are doing part of their job or you persuade them that what you're doing is good for the job as a whole. Architects could use engineers who are cheaper; they come to someone like me because they are looking for a certain level of professionalism and quality.

TUCHMAN: I get the sense that you feel exposing or expressing structure is a good thing....

LEVY: I do, very much so. I still feel that structural expression is not only pleasing, it's also highly economical. It has secondary benefits. For example, when a structure is exposed, you can always see what's happening to it...if it's behaving well. Though it's certainly not for every building.

TOMASETTI: What is the most technically difficult problem in exposing structure, using the definition of a structure which is actually exposed to the environment?

LEVY: The details and the proportion of the details...how a particular joint is worked out...how you bring the beams into the column...how you satisfy the need for two different kinds of joints, some fixed, some sliding. These details are what the structure will be judged by.

TUCHMAN: What about working with materials other than steel or concrete?

LEVY: One thing I thought was interesting recently is the work a Swiss engineer is doing with wood structures. He's been developing a lot of fun joints for wood, very expressive joints for the trusses and beams. I think it has brought back a kind of structural romanticism to designing wood structures.

TUCHMAN: Do you have to be careful with wood?

LEVY: Wood joints are very critical. They are hard to design because wood has characteristics unusual in a structural material. Unlike steel, wood doesn't have the same properties in all directions. Plus, you can't weld it, and even when you bolt it, you have to be careful...it's very tricky (Chap. 6).

WALTER P. MOORE

Houston-based structural engineer Walter P. Moore shares the feeling of many of his colleagues—he likes to see his structural frames exposed before the cladding is added, and he often photographs them. He cites fire protection concerns as the major reason steel structures are not frequently exposed.

Moore was involved in the design of one of the tallest cast-in-place exposed concrete structures in the United States, the 1-million-square-foot, 40-story Barnett Plaza in Fort Worth, Texas. The architects were Geren Associates/CRSS and Sikes, Jennings, Kelly, and Brewer. Floor framing is haunched girders and pan joints. Lateral loads are resisted by a perimeter tube system of closely spaced columns and deep spandrels, working with core shear tubes. Thermal effects on the structure were specially analyzed because the architecturally exposed concrete in this system experiences large temperature swings.

MOORE: One interesting exposed structure we did was in Fort Worth, completed in 1983. It was a 40-story building with punched windows in a total concrete frame with columns and girders. The frame became the exterior facade, while the window wall was just glass infill between the columns and the girders (Fig. 3.26).

THORNTON: The outside face of the columns and girders was left exposed?

MOORE: We actually put slip joints in the exterior spandrel beams to let some thermal movement occur, because you wind up with such stiff columns and girders.

THORNTON: How did that work?

MOORE: In each girder we had two tubes that telescoped inside adjacent tubes at the expansion joints. They functioned as expansion joints, which moved in and out, so the structure could breathe and relieve the buildup of thermal stresses. These spandrel beams were exposed on all four sides.

THORNTON: Do you remember the first couple of meetings on that project? There must have been an architect, a structural engineer, a construction manager, right? Who pushed for the concept that was ultimately selected, the exposed structural frame? Why did it happen?

MOORE: The major driving force was that they wanted the building done immediately and cheaply. It was at the height of the Texas building boom and they actually added 10 floors after it was under construction. It was one of those projects where you're told: "Just get started, don't worry about a thing."

THORNTON: Fee was no object?

MOORE: Fee was no object—just get started and we'll work out these details as we go along. Well, they added 10 floors. It had to be an inexpensive building. The facade had to match an adjacent existing building, which was clad with precast concrete, and architectural concrete seemed like a natural choice. So an idea came up, why don't we just leave the structure exposed? The contractor resisted it like crazy, because he was scared of the problems of producing an aesthetically acceptable result. But in the end he succumbed. Since the owner liked the idea, the architect pushed it. In that one case, I didn't push so much as respond to it. I was a little nervous about it, too.

Figure 3.26 The fine architectural finish on this building is simply the cast-in-place concrete structural frame.
Barnett Plaza, Fort Worth, Tex.; Architects: *Geren Assoc./CRSS; Sikes, Jennings, Kelly, and Brewer*; Engineer: *The Datum/Moore Partnership.* (Photo: *Walter P. Moore and Associates.*)

THORNTON: Was it the architect's idea or the owner's?

MOORE: You know, you think back to these meetings, say six or eight years ago, and you forget whose idea it was. Most ideas that are any good are joint ideas. Someone says something, another guy on the team picks up on it, and you sort of arrive at these things. Then everyone takes credit for it. And everyone does have a little part in it.

THORNTON: I can imagine that what happened is, you evolved a structural solution which eliminated the skin. You still had to do windows, but you could buy individual windows more easily than you could an entire skin. And there are times when building booms hit, and you can't even get the skin. So it was probably the owner who said, "Hey, this is a great idea because we don't need a skin!"

MOORE: And consider what it cost per square foot. Lots of times your curtain wall system costs more than the entire structure. The contractor verified that we were saving a lot of money by using this system even though the cost of the structure was higher.

THORNTON: In this case, you were supplying not only the structural framing, but also the skin.

MOORE: You don't get any credit for that, either. And there's no add-on to your fee per square foot.

THORNTON: No one told you when you negotiated your fee that this was going to be an exposed skin structure. Generally, whether a structure is exposed or unexposed is unknown at that point. In your work, does exposing or expressing the structure take more design hours, a larger work force, than concealing it?

MOORE: Oh sure, if it's the main structural elements you're talking about. In this case, with the columns half inside and half outside, you've really got to wonder about the effect of temperature gradient. It takes a lot more time and a lot more attention. And then you have to consider the detailing and the joints (Fig. 3.27). So when you take on the exposed structure, you spend a lot more time. It's much easier to cover up a structure with fireproofing and go on to the next job.

THORNTON: And there's no consideration of it in the fee discussion in the early stages of the project?

MOORE: No one would quote a fee like engineers quote a fee. "I've got a million-square-foot building, Charlie, what would you do it for? I'm not going to tell you what it looks like, I'm not going to tell you what it's made of...."

THORNTON: "I'm not going to tell you what the foundations are... ."

MOORE: "Nothing—just tell me what your fee is." We're all lunatics, but we all do it.

Figure 3.27 Cantilevers and rustication joints are of exposed structural concrete. Tie-hole patterns add to the carefully detailed effect.
Barnett Plaza, Fort Worth, Tex.; Architects: *Geren Assoc./CRSS; Sikes, Jennings, Kelly, and Brewer*; Engineer: *The Datum/Moore Partnership.* (Photo: *Walter P. Moore and Associates.*)

THORNTON: I only do it because you're doing it. When I go into a meeting, they say, "Walter said he could do it without this information." But architects do it, too.

MOORE: Yes, but contractors don't.

THORNTON: Well, they're smarter. They give it as a percentage of the construction fee, at least.

MOORE: Anyway, it takes a much greater effort to plan and respond to exposed structure. For example, on another project we're looking into some sort of coating or paint that would be acceptable for fireproofing. We're researching something that's not normally part of our structural design.

THORNTON: You mean you find yourself doing work the architect would handle when designing the cladding? In other words, it no longer holds that if it's nonstructural the architect does it, and if it's structural the engineer does it.

MOORE: There's more and more added to the engineer's problems.

THORNTON: Have your fees gone up as a result of that?

MOORE: They have not. The owners are trying to make the curtain wall more our responsibility, so now we're fighting back by saying, "Let's see an engineer's seal on it; we want to get paid to do that work." That draws a lot of griping, but they're beginning to fall into line.

THORNTON: I think that has merit in terms of safety.

MOORE: It does have merit, and that's why it's basically been accepted. It's one of the few times all the structural engineers have seemed to stick together; no one has let the owners off the hook on that.

THORNTON: Think about the number of times you've been in a meeting: the architect has fallen in love with an exposed structural concept. He's convinced you to fall in love with it. But you both know it's going to cost more, not only to build, but for you to design. Still, you want to maintain the creative image of Walter Moore, so you don't say, "I can't do that." You're now in cahoots with the architect. Everybody is trying to convince everybody else that this is the way to go.

The owner and the construction manager come in and the politicking starts. But in the back of your mind you're praying that some philistine will arrive, blow the whistle, and put a stop to it. Sometimes the scheme, which you know in your heart is going to cost more, gets killed, and you go on to something more within the scope of the project. And sometimes it doesn't. This is where the politics comes into the process.

MOORE: It's fine for the architect and me to get together and decide that this is what we want. But the next question is: who has to be sold first? There's a construction manager who's apt to come in and say right off the bat, you can't afford to do this. If, on the other hand, the owner has been sold on the concept first, and his ego gets into it, then you're going to go merrily ahead while they complain about the cost of the building from that day forward.

THORNTON: Are you saying ego plays a big part in this?

MOORE: Ego plays a tremendous role—ego and pride.

THORNTON: Do you think exposed structures are more daring?

MOORE: I really don't know. Some concealed structural projects are very daring. And some exposed structures are very straightforward and obvious. But certainly when you expose a structure you have a sense of daring. Now, the building in Fort Worth I mentioned earlier will not look daring to you—you'll say that it's an ordinary building. But you don't see all that went into it. From my standpoint, it was very daring.

THORNTON: I'd like to ask you about parking structures.

MOORE: Basically, that's the ultimate exposed structure.

THORNTON: They're not exciting or heroic. You don't do them on an ego trip. The fact of the matter is they frequently aren't designed correctly, and because they're exposed, they don't last very long.

MOORE: You're right, they don't. They're a serious problem, like the infrastructure of the cities. There are millions and millions of square feet of garages on the verge of falling down...after a very short life span.

THORNTON: Given two posttensioned concrete designs, would you do it differently in Dallas or Houston versus Chicago? Is an exposed concrete frame, which has a tendency to develop some shrinkage cracking or water intrusion, less of a problem in a climate where it doesn't go below freezing?

MOORE: I really don't think the temperature's the problem. What I'd be more scared of is the exposed structure on the east coast of Florida, say, somewhere where you're going to be subjected to all that salt water. Those buildings really take a beating. But probably in Houston we would have a little more control over the mixes, and the concrete would be more durable. Now, we've changed our philosophy on garages a great deal in Texas because of all the problems. It used to be that for so many square feet of office space it was assumed you had to have so many square feet of parking space. No one paid any attention to garages. One developer, when we told him the fee for his office building, figured it included the garage. It was supposed to be done very cheap. They were all going precast with no control over the precasters. They went up so fast, most serious people didn't realize the rules that were being broken. We had one collapse because the tendons weren't put in properly. There were no stirrups in the girders. But we weren't smart enough or paid enough to check up on the precasters like we do now.

THORNTON: Did the average structural engineer on a large project, for the most part, leave the garage alone?

MOORE: We did nothing. We let the precaster do it. I'm not sure we even did the calculations. We were hired to do the foundations. The precaster gave us the loads to put on the footings. We didn't have to show anything. Frankly, we didn't even know what the garage would look like. This went on all over.

We had one funny situation where we were doing a basement garage with a very unusual precast, posttensioned wall system, where the tendons kind of go around curves. The owner asked me to take a look and tell him if it was any good. So I called up the guy, and I was told it was secret. I reported back that the design of the wall was secret. The owner said, "What do you think about it?" I said, "I don't have any thoughts about it other than that it's secret." Well, it subsequently failed a year later.

THORNTON: But I'm sure, in those days, if you were the base building engineer of record and the other guy wasn't a P.E., the problems could have been attributed to you. If the detail or technique is secret, then the one who designed it ought to be damn well responsible for it.

MOORE: Sure, it wasn't secret, it was stupid. And there have been a lot of very costly problems. We're ripping one garage out and doing it over right now.

THORNTON: With precast, do you find that volumetric changes—the thermal, the creep, and shrinkage—cause spalling and cracking?

MOORE: All the joints get water in them. Precast doesn't wear well. The garage doesn't get maintained, everything's rusting away, and the rust is holding up the whole thing. Plus, the water drips through all the stuff. Some of the slabs are topped, but the topping cracks at every joint—no mesh in it. Even with mesh, they crack.

If you just stand in one of these things, just let your car run down the main driving aisle and feel the structure, you know something's got to give. I'll bet you the life span of these garages will be very short. I don't know how they perform up here.

THORNTON: The history in the northeast is that a couple of companies really pushed the precast tees and convinced a lot of clients they were...

MOORE: ...cheap, cheap, cheap.

THORNTON: But also durable, durable, durable.

MOORE: No, not durable, durable; cheap, cheap. We're discouraging precast garages, we've had it up to here with them.

THORNTON: In favor of what?

MOORE: Posttensioned, poured in place. Because it's monolithic, it's just a much better system. You pay attention to temperature... you design it like you assumed someone was doing all along with the precast. But that was not the way it was; it was just a series of toys.

You know, the ACI is adding integrity provisions to the new code. The precast people were very suspicious that this was an attempt to get them. It is a chance to make them tie these structures together.

THORNTON: Of course, the parking structures are just a small part of a large contract. The contractors are generally small-time, and they come and go. The same way with facades. Some guy comes out with a new miracle exposed aggregate panel; the architect says, I wouldn't use it if I were you; the owner says I want it, it's cheap. The next job, it's another company—they just keep starting up and stopping.

MOORE: We also no longer do very much tilt-up. I think it has a lot of the same problems; no one pays attention to anything, except that the contractor who's lifting up the panel has to take some care so it doesn't break.

THORNTON: In the classic tilt-up warehouse, does the exterior wall support the end bay, or is it basically tilted up against a light steel frame?

MOORE: Typically, it supports the last bay of the roof. And it theoretically supplies the stiffness for the wind, which it does if you can rationalize a diaphragm at roof level—which I think is highly questionable in lots of situations.

THORNTON: Especially if it's not engineered. We've redesigned one in New Jersey, 700,000 square feet. It was so big there were a couple of bays in the middle that were totally isolated from the exterior wall. When the wind blew on the building, one part banged against the other.

MOORE: I think the problem of buildings that are not engineered at all is much more rampant than we think. Engineers respond by making up all these rules that tie our hands even more. Yet they have no impact on the guys building the dangerous structures. They're just running loose. That's what I object to. But it's getting better. There are many malpractice cases. It's not just engineers speaking out. It's others getting involved in the process.

PETER RICE

A discussion with Peter Rice, Director of Ove Arup & Partners in London, was a must for a book on exposed structure because he has collaborated with architects Richard Rogers and Renzo Piano on some of the landmark exposed structures of recent years. The three designed Centre Pompidou in Paris, and with Rogers, Rice designed Lloyd's of London (see Figs. 2.31 and 3.28). But Rice believes it is a mistake to think there is anything unusual about using structure as an architectural medium. He says it is part of a highly traditional approach to building.

TUCHMAN: Do you feel that an exposed structure adds excitement or daring to a project?

RICE: You have to be careful not to imply that there's something

Figure 3.28 Exposed concrete column/bracket/bracing joint showing excellent forming placing and finishing.
Lloyd's of London, London, England; Architect: *Richard Rogers Partnership*; Engineer: *Ove Arup & Partners*. (Photo: *Ove Arup & Partners*.)

unusual about exposed structures. They've existed as part of the framework of buildings for a very long time. If you explore Paris, for instance, you'll find lots of 19th century iron and steel buildings with completely exposed structures. Admittedly, they represent a tradition which predates the understanding of fire protection, but they are technically solid and aesthetic frameworks.

TUCHMAN: Would you elaborate on the historic aspect?

RICE: Yes. The very first evidence of a structure is in stone. Building forms derived from the kind of structure you could erect in stone, then in brick, then using other kinds of non-tension-carrying materials. Gradually, as technology improved, people were able to modify and change the way the structural behavior of materials worked. Still, the forms that were developed were based on the structural properties of stone, which, as part of a rich history, became associated with classical structural forms. But it's a mistake to think there is anything unusual about using structure as a medium.

If we talk about the use of materials such as steel, all that my colleagues and I have been trying to do is define an aesthetic based on the properties and characteristics of steel. This is an aesthetic which was fairly thoroughly explored in the 1800s. In this century it was developed in a variety of ways by people like Frank Lloyd Wright, Louis Kahn, and Mies van der Rohe. Many of these people have taken structure as an expression of material behavior and used it as the aesthetic element that defines the way they approach architecture. So I don't think it's fair—or even illuminating—to consider ourselves in any kind of special situation today.

TUCHMAN: But if you look at all buildings built in the 1980s, only a fraction of a percent have exposed structures.

RICE: No, what I mean is that their aesthetic expressions describe or lean on structural expressions from previous periods. Take, for example, the classic tall tower. In many ways it derives from—or is a multiplication of—thinking used by people like Mies van der Rohe. It's Mies's fundamental position, an aesthetic based on a structural frame.

TUCHMAN: But people are no longer always choosing to have a direct relationship between what appears on the facade of the building and the structure needed to hold it up.

RICE: I would agree with you. We are now using classical and traditional forms as if based on the structural properties of materials like brick and stone, but we are actually building them in a different way.

TUCHMAN: Using different materials.

RICE: That kind of subtle transition, that difference in character, is probably characteristic of everything we do today. Even when we're talking about exposed steel structures, as in Centre Pompidou, or exposed concrete, as in the Sydney Opera House (see Fig. 1.8), you're now following simple structural logic and composition. You're using a structural material, or elements of its properties, to give aesthetic value to a structural approach. But in no way are those buildings representative of fundamental structural approaches. They're using the structure as part of the aesthetic framework in much the same way people did with brick and stone in the 18th and 19th centuries. Now we use steel and concrete, and it's inevitable that one of the ways of architecturally expressing the character of those materials is to let their use as part of the structure be evident, even though it may not be a logical structure in the first place.

TUCHMAN: Can a structure be illogical?

RICE: Of course it can, though there is no single definition of what is logical. It depends on one's prejudices and the quality of materials on hand. What appears logical to an American engineer may be the converse of what's logical to me, since we live in different environments. So I take the view that the use of structure as an exposed element is almost an architectural decision, part of an architectural philosophy that requires a degree of dialogue between engineer and architect. It's not necessarily an extended dialogue, or one that is particularly sophisticated, because sophisticated structural concepts don't usually make good architecture. I have a rough principle that I use in design of structure. Never do in one joint what you can do in two. It's difficult to read the performance of a joint that's doing two jobs. Exposed structure as an architectural element is all part of a highly traditional approach to building. It may not be part of the short-term American tradition, but it might be part of the long-term tradition.

TUCHMAN: I think perhaps recently we've seen fewer new structural theories or great leaps of understanding of new structural systems. Is that affecting the use of exposed structure?

RICE: I don't think that what we're doing when we expose structure today is more—to use a behavioral term—complex. I think what is different today is the sophistication. Remember, that's a two-edged word that also means sophistry, or arguing cases without foundation, and that implies a degree of trickery and sleight of hand. What is evident is that the rules by which buildings are designed, or the rules they have to satisfy, have become progressively more stringent and strained in terms of fire, structural safety, and collapse. This makes the use of structure as a primary means of expression much more difficult to achieve, but I don't think they're more complex than they would have been a hundred years ago. For instance, nothing is being built today that's as visually and technically complex as the Arc de Triomphe.

TUCHMAN: Nothing as intricate.

RICE: As intricate and as subtle in the way it minimizes the use of material. It's every bit as sophisticated as anything we do today.

TUCHMAN: Maybe I'd gain better understanding of your point of view if we talked about some of your projects.

RICE: The Centre Pompidou is a very good case in point. It illustrates some of the problems quite clearly. It's conceived as a Greek-style building in the center of Paris, as a framework within which other things happen. That framework was an intricately conceived steel structure, designed to achieve a variety of different things. Since it was in the heart of Paris, we were particularly concerned that it have a level of subtlety and complexity that would not be perceived as aggressive (see Fig. 2.30). There is another important aspect to this kind of framework. While exposed structure is an unusual architectural element in buildings, exposed structure per se is very usual. It's all over the place, in bridges, scaffolding, factories, frameworks for towers. In New York, you see it on top of all the buildings holding up the cooling towers.

Because of that, when people approach any building that incorporates exposed structure, they use a built-in set of references

Figure 3.29   The truss joints shown here were made of castings rather than more conventional weldments to emphasize the difference between this building and the industrial vernacular.
Centre Pompidou/Beaubourg, Paris, France; Architects: *Piano+Rogers*; Engineer: *Ove Arup & Partners*. (Photo: *Richard Einzig/Arcaid*.)

as a means of understanding what they're seeing. That produces a kind of prejudice, and the danger is that an exposed structure will be perceived as something else, not as a building. People have used the term "oil refinery" when discussing the Centre Pompidou. That's the kind of industrial image that becomes associated with it, and we were very anxious to break that connection. So we decided to use cast steel, which hasn't been used in the building environment for 70 or 75 years. It's used in boats, in machinery, but it's considered unacceptable in structures, due to the flaws that can develop during the manufacturing process and may precipitate failure. That kind of arbitrary behavior is unacceptable, so in deciding to use cast steel we had to develop an approach to specifications that gave it the same level of public and code performance as normally produced steel systems.

TUCHMAN: When you say we, you mean...

RICE: My design team and I. The safety aspect was obviously our top concern, and we decided we had to take the technological approach of fracture mechanics. We developed a theory for cast steel that took into account the size of flaws that could be detected, the scale and size of the pieces, the way they were cut, the way they were treated, so that in effect we could guarantee the same level of confidence to be expected of materials produced by

Figure 3.30 Cast steel arms are clearly visible here.
Centre Pompidou/Beaubourg, Paris, France; Architects: *Piano+Rogers*; Engineer: *Ove Arup & Partners*. (Photo: *Ove Arup & Partners*.)

other means. In a sense, we had to adopt the very latest in 19th century industry to achieve the way it was made 50 or 60 years ago. But the way we specified, the way the relationship between the manufacturer and the process and the performance of the material was defined, the kind of testing we did to back it up, all were completely different. We gave ourselves a freedom to design pieces that would have been normal 70 years ago. We did this because we wanted to break the perception in people's minds that the structure was part of some other image they were carrying around with them (Figs. 3.29 and 3.30).

TUCHMAN: That was your decision, or was that something the architects...

RICE: Ours. The whole steel structure was really designed by the engineers, with the architects working with us in an architectural framework.

TUCHMAN: Was this a joint venture?

RICE: I worked on it as part of an engineering team, and obviously there was an architectural team, three people working together. But the choice of materials, the design of the steel structure, and the way it was developed were clearly an engineering decision that had to be accepted architecturally. One person doesn't do something any of the others disagree with. The whole thing became an exploration, but the decisions to do certain things had to evolve, like the responsibility of the engineer for the components—which was key. An architect couldn't propose to use cast steel in that situation then, because the material had to be made to work, and that had to be done by an engineer.

LESLIE E. ROBERTSON

Leslie E. Robertson, partner in the New York City–based structural engineering firm Leslie E. Robertson Associates, is best known for his pioneering wind engineering research and for the structural design of very tall buildings, such as the 1368- and 1362-foot-tall World Trade Center towers in New York City and the 1209-foot-tall Bank of China tower in Hong Kong. He says all three buildings express their structure, but the Bank of China says it more forcefully, appropriate to the heavy wind loads it must resist. He believes expressed and exposed structures are like any others in that the architect remains key to the formation of the design. Structural engineers and others contribute ideas, but the architect chooses among them to create the vision of the building. Robertson engineered one of the country's exposed structural landmarks: the 3-million-square-foot U.S. Steel Building in Pittsburgh, Pennsylvania. Innovations there included liquid-filled columns to provide fire resistance.

THORNTON: Why did you use expressed structure for the Bank of China and not, for example, on the World Trade Center?

ROBERTSON: Actually, I wouldn't agree with that. It seems to me that when we talk about structure that is expressed, we have to talk about people's perception of it. I think most people accurately perceive the World Trade Center for what it is. The structures that are really deceiving, those with false columns and the like, go back to the days when we used induction units, when we had those round, high-velocity ducts. Even the last buildings of Rockefeller Center have fake columns. If there is an expressed structure in cases such as these, you can see it where there are columns; but I think the public sees columns where there are no columns. On the Trade Center, however, they see columns where there are columns. So, in a sense, the Trade Center is more of a sculptural form than an architectural form—it reads almost as a solid surface in many ways (Fig. 3.31).

TUCHMAN: So you feel that the World Trade Center is an expressed structure?

ROBERTSON: Yes. It's no different than the Bank of China, except that the Bank of China says its message more clearly. It says, "I'm structure" rather forcefully, and in a place like Hong Kong that's important, because everyone who lives and works in Hong Kong understands that the wind loads are enormous and that this structure is significantly taller than anything that's been built in that part of the world.

TUCHMAN: But the structure is selectively expressed. Parts of it are hidden (Figs. 3.32 and 3.33).

ROBERTSON: That's true of all structures; you can't express everything. The Bank of China tower was difficult because the architect, I. M. Pei, had an approach to the design that made use of the secondary, horizontal trusses. He had attached an importance to them that went far beyond what I attached to them; to me they were a secondary issue, dealing with the way in which the curtain wall was organized; to me they dealt with the needs of vertical wind loads between panel points of the great megastructure. I. M. Pei, on the other hand, saw them much more as the physical transfer of load. There was a lot of philosophical groping with that issue.

Figure 3.31 Columns for the perimeter tube frames of these twin towers are closely spaced and create an impression of texture, not framing, at a distance.
World Trade Center 1 and 2, New York, N.Y.; Architect: *Minoru Yamasaki with Emery Roth & Sons*; Engineer: *Skilling Helle Christiansen & Robertson*. (Photo: *Jerry Rosen, Port Authority of New York and New Jersey.*)

Figure 3.32 Glazing the areas between braces creates diamonds, an auspicious symbol here.
Bank of China, Hong Kong; Architect: *I. M. Pei and Partners*; Engineer: *Leslie E. Robertson Associates.* (Photo: *John Nye, courtesy of Pei Cobb Freed & Partners.*)

Figure 3.33 Structure is dramatically exposed in this penthouse corner, reminding visitors of the scheme expressed outside.
Bank of China, Hong Kong; Architect: *I. M. Pei and Partners*; Engineer: *Leslie E. Robertson Associates.* (Photo: *Paul Warchol.*)

THORNTON: Wasn't there a problem with exposing or expressing the horizontal load because it created a symbol that was not positive in the Chinese culture?

ROBERTSON: That was another issue. By incorporating the horizontal bands expressing the trusses at the top and bottom of the 13-floor increments, the huge expressed diagonals that brace the tower were perceived as Xs, a negative symbol in the Chinese culture. I believe the bank's posture at the time was that they thought the structure was great and they wanted to build it just as it was, except that they wanted to put an ordinary curtain wall on it and to omit any expression of the structure. But I. M. is not the sort of person you push around. He has his ideas, and if he can't reconcile them, he's not going to build. The question of bracing the 13th floor just reminded him of all the time the two of us and others had spent trying to arrive at an understanding of the importance of that particular part of the structure. My view was that the addition of the trusses was realistic in terms of making the structural work; but in my view the building is handsomer and simpler without the expression of the secondary trusses. Eventually I. M. elected to redesign the facade. By leaving out the horizontals, the pattern becomes a diamond, an auspicious symbol (see Fig. 1.1).

TUCHMAN: How do you feel about the process that takes place between the engineer and the architect, especially at the point at which you're deciding whether the structure is going to be expressed or exposed? How does that give-and-take work?

ROBERTSON: In my experience it really doesn't. It's the sort of thing you arrive at instantly on a project. There's a design, and you either have the structure there or you don't. Either you have the high-velocity ducts expressed or you don't. It's a part of the concept; I don't think it's debated.

THORNTON: What Jan is alluding to is that everybody is in quest of the perfect building, and the perfect building would be one in which you could expose the whole structure. If you've got these great structural shapes and configurations, why waste money covering them? Why not go with an exposed structure?

ROBERTSON: Waste money?

THORNTON: We're trying to get at the question of whether it is false economy to think of exposing the structure and eliminating the skin. You did the U.S. Steel Building in Pittsburgh; it was a totally exposed steel structure (Fig. 3.34).

ROBERTSON: It has an exposed structure, but it's not totally exposed. You feel you understand the structure on the Bank of

Figure 3.34 This showcase of steel has unpainted surfaces, liquid-filled columns for fire protection, and a megaframe which carries stacks of several floors at a time. **U.S. Steel Building, Pittsburgh, Pa.**; Architect: *Harrison and Abramowitz*. Engineer: *Skilling Helle Christiansen and Robertson*. (Photo: *Leslie E. Robertson Associates*.)

China, but on the U.S. Steel Building, while you see the actual structure and while you may know it's exposed steel, I don't think you understand how the structure works.

THORNTON: Do you think it's important for a person, whether he is an architect, an engineer, or a layperson, to really understand the structure?

ROBERTSON: I think it's important that everyone feel comfortable with the structure. I don't think it's important to understand it. For example, the RCA Building or the Seagram Building are fine buildings. Do you understand the structure? I don't think so. No real attempt was made to express the structure, and I don't see anything wrong with that.

THORNTON: How about Edward Larrabee Barnes's IBM Building on the corner of 57th Street and Madison Avenue?

ROBERTSON: Now that is more controversial because it brings up the question of being comfortable with the building (see Fig. 1.5).

THORNTON: Because of the dramatic cantilever overhanging the entrance?

ROBERTSON: Yes. The building he did just a few blocks away that adds a column in the corner probably makes people feel comfortable (see Fig. 1.6).

THORNTON: Expressing the structure may create drama and thus discomfort. Or, going back to the Bank of China, the expressed structure lets people know that this building is going to be stable against the strong winds.

ROBERTSON: Yes. I think that's a good reason to do it.

TUCHMAN: What we seem to be saying is that the expressed structure gives the public a sense of how the structure performs. They can read the building. But exposing the material might be done for a different reason. Why is material being exposed to the air?

ROBERTSON: If we take the U.S. Steel Building as an example, it's because it is the U.S. Steel Building. In that case, there was a driving force to meet the commercial needs of the builder.

TUCHMAN: On the other extreme, however, the material in a concrete parking garage is exposed for the purpose of saving money.

THORNTON: There are some buildings in which the structure and the mechanical systems are all on the outside. The theory is that because the mechanical systems keep changing and the structure never changes, you hang everything on the outside and keep the interior of the building totally free for functional changes. Our experience, however, is that this costs a fortune. Do you have any comments on that?

ROBERTSON: I think it's nice theory, but I don't think it's a theory that's going to save money. I think it's going to make it very expensive to change the mechanical systems in the future because the mechanical system is very delicately hung outside. It's not going to be easy to make changes without also affecting the architecture.

TUCHMAN: This is what I was getting at in my other question. Where do the ideas come from? Isn't there a give-and-take as the building is being designed?

ROBERTSON: The architect is the designer of the building. People sit around and talk about things, so I don't think ideas are very assignable to individuals. Everyone is part of the process. However, to say that the Bank of China is anyone's design other than I. M.'s is stupid. Let me assure you on that; while we contributed much, it was completely his design. He used whatever he felt like using, and he discarded whatever he didn't want. Out of that came a very fine building because he is a very fine architect. All the ideas can be there, but so what? You still need that great designer. That's what makes the world turn in our business.

THORNTON: Do you have any comments on what exposing a building does to the engineering—the total hours, the fees, the effort, and the detailing?

ROBERTSON: First of all, when you construct a building, it's not in an air-conditioned environment, which means that for some amount of time the building is thermally exposed. When you start looking at the return period of high and low temperatures and at the joint probabilities of either high or low temperatures and winds or earthquakes, you find that you can almost isolate the two. But I don't think that exposing the structure creates a particularly difficult structural problem because you've got to design for these extremes anyway when you're building the building. It does mean that you have to be a lot more careful in the way you detail things since they affect the parts that grow and shrink. For example, if you made rigid or nonductile connections to things, they could very well break.

THORNTON: One of the things that has always amazed me about Chicago is that there are all of these 30-, 40-, 50-, and 60-story concrete buildings with just paint on them. I've been involved in some projects in which the partitions that went from the core to the outside were jammed in so tight that you got cracking in the partitions even on a 10-story building. Have you ever done any tall, exposed concrete buildings like that?

ROBERTSON: Most of what we have done is exposed. The tallest one is in Hong Kong, 65 stories, and it's all exposed.

THORNTON: What input did you have on the Hong Kong building in terms of making sure that architecturally the other subsystems were designed to take motion?

ROBERTSON: We just provide the deformations to the architect. With concrete buildings the long-term effects were quite significant. The effect of temperature is less severe. For example, for the U.S. Steel Building we did a study at the Chase Manhattan Bank Building in Manhattan. It has columns on the outside and they're clad and insulated so there is a thermal lag, although there are still significant temperature changes. In the winter those columns get cold right down to the ambient temperature. We ran levels in that building to see what kind of motions you could tolerate on the outside wall and still get the partitions and doors to open and close. That's the way we arrived at the criteria for U.S. Steel.

THORNTON: I mention this because so many engineers who work on 5- and 10-story buildings never even discuss such criteria for motion with the architect.

ROBERTSON: But this is one facet of a whole series of volumetric changes that take place in a building which you have to report to the architect if you want to keep your skirts clean. It's not just reporting it; you have to communicate it to the architect so that he understands it.

Less communication is necessary if the building is inside the experience of the builders. The best example is a simple shaft wall. That was the idea we came up with for the World Trade Center. That system then got applied to many other buildings that would have been stabilized laterally by masonry partitions but suddenly didn't have them anymore. Now, instead of masonry, they had this lightweight partition, and the stiffness went way down. In a sense, the problem could be laid at my doorstep because I didn't write a paper telling engineers to start worrying about mass because they were using this partition system. We had designed the Trade Center with Teflon joints in the partitions and artificial dampers in the structural system. But on the next buildings down the line, some architects and engineers didn't realize what happened to them.

TUCHMAN: I'd like to go back to the question of cost. Are exposed structures more expensive or less?

ROBERTSON: Just putting up a plain repetitious curtain wall is going to be less expensive than exposing structure or expressing it.

There's no question about it; you're going to save money.

On the other hand, that may not necessarily follow. Because for the same level of architectural expression, something must be provided to take the place of the exposed structure. I don't think anybody really knows whether you save money or not. But to do it just to save money is unjustified.

THORNTON: I think the only structure for which it probably does save money is the exposed concrete residential building.

ROBERTSON: Or the parking garage.

THORNTON: Moving to an interior situation, how many projects have you worked on in which an architect says, "I want to integrate the mechanical, electrical, and structural. I want to expose it all." Then the budget comes in double what it would be for a normal building?

ROBERTSON: It happens all the time. You get a building completely designed with all the exposed mechanical systems, and you just can't afford it.

THORNTON: Let's talk a little bit about fireproofing. I was curious whether you have any thoughts about the evolving trends with exposed steel and trying to get away from fireproofing.

ROBERTSON: You can get away from it by liquid filling, as we did at U.S. Steel; that's expensive. And the German DIN standards that allow the flanges to be exposed seem to be a completely rational, sensible way to go about it. In England, too, there are prefabricated fireproofing sections available. I think we could stand to move in that direction. There are a lot of instances in which we are adding fireproofing in this environment when we don't have to. It's very expensive and adds to the cost of maintenance.

THORNTON: That's what BHP, the major steel company in Australia, is saying. And they've got full-scale fire tests to prove it. I think we'll see some innovative things in this area.

You said before that engineers have to communicate with architects. But architects and engineers have to communicate with contractors—that's the second step that many times gets overlooked.

TUCHMAN: Even down to the level of the actual workers.

THORNTON: When we did that stress skin hangar in California with the light-gauge metal roof, we had a seminar with all the welders to explain to them that this was not just a typical roof decking— this was structure. They hadn't realized that all these panels were going to be used that way (see Fig. 8.14).

ROBERTSON: Absolutely. If something is important, the engineer has to go and say, "Hey, look at this." It's not enough to write it down on the specification, to note it in the drawings, or to write it for the testing agency. You've got to go out and make sure it's being done.

LORING A. WYLLIE

Loring A. Wyllie, Jr., Chairman of San Francisco-based consulting engineer H. J. Degenkolb Associates, did not immediately tend to think of his designs in terms of exposed structure. Foremost in his mind, as a seismic design specialist, was designing for earthquake resistance.

But as Wyllie warmed to the subject, he thought of a number of examples of exposed or expressed structures in his work. And he pointed out that the most common use of exposed structures in earthquake-prone areas is for strengthening damaged structures or bringing underdesigned buildings up to code. Additional bracing can be added to the outside of a building without taking it out of service.

Wyllie wrote to us after this interview, for example, about a seismic retrofit of University Hall, the Administration Building at the University of California at Berkeley. This was a concrete frame building designed in the 1950s, and with deep 8-inch-thick spandrel beams it was a strong beam/weak nonductile column frame, one of the most vulnerable types of structures in a seismic environment. The exposed steel diagonal bracing was designed and constructed after the 1989 Loma Prieta earthquake (Fig. 3.35). The scheme was more economical than other solutions, as few finishes were disturbed. Being an exterior scheme, it also allowed the building to remain occupied while the work was performed. Wyllie says he has had several people, including architects, compliment him on how it improved the appearance of the building. In the University review process he adds, they never complained about the appearance of the braces, only the color the architect originally proposed to paint them. "We were very careful about our connections and details, going to difficult full-penetration weld details to give clean lines and avoid ugly gusset plates, which can be more economical. When we expose our structures, we need to be sensitive to these issues of aesthetics," he said.

TOMASETTI: In our discussions with architects and engineers, one type of exposed structure that keeps cropping up is the parking

Figure 3.35 Photos show the nature of exposed retrofitted seismic braces on this plain rectangular building. While not a reason for the work, the braces add visual interest. **Retrofit of University Hall, UC Berkeley, Calif.**; Architect:*Hansen/Murakami/Eshima*; Engineer: *H. J. Degenkolb Associates*. (Photos: *Mario A. Fovinci.)*

garage. As someone who works predominantly on the West Coast, how would you say your concerns are different when it comes to handling garages?

WYLLIE: We don't have the same corrosive elements in California that you have here—the salts and all that—unless you're right on the coast. Still, it's amazing how many parking garages need fixing up.

TOMASETTI: Of course, everything you do has to take earthquakes into account.

WYLLIE: That's right. One problem we have that you don't, is adding shear walls to these buildings. We see a lot of either precast or posttensioned garages without shear walls or rigid restraint. You get so much creep and shrinkage in those unbonded slabs that a tug of war sets in. But if the restraints aren't placed just right, the structures simply pull themselves apart (see Chap. 7).

TOMASETTI: The term "exposed structure" means different things to different people. In light of what we're discussing right now, how do you interpret it?

WYLLIE: Certainly some of the shear walls in buildings are exposed, but they are just walls. I don't necessarily consider that an example of exposed structure, per se. But in a garage the structure is almost always exposed, I guess, so, sure, it qualifies. I tend to think more in terms of the Hancock building in Chicago (see Fig. 3.13) or the Alcoa building in San Francisco, where even though the structures are covered in cladding, they're made a feature of the architecture. For some reason we don't seem to have a lot of those buildings in California.

TOMASETTI: Do you feel the code requirements for designing connections to resist earthquake forces have an effect on the more exposed structures?

WYLLIE: I wonder about that. I think the bigger buildings are mostly steel-framed, and there are some exposed steel buildings. But those are the kind you tend to wrap up, to hide the steel a little more. So I tend to see more conventional buildings, wrapped up with the structure hidden. However, with the increased use of diagonal bracing, I think a little more "exposure" is coming about on the West Coast. I think we're beginning to see a few diagonal braces here and there, because diagonal bracing is a lot better than a shear wall. You can see through it. You get a diagonal to contend with, but if you work with an architect who's got an open mind, you can usually work up a compatible system that turns out pretty well.

TOMASETTI: But you do see some exposed structure on the West Coast. Was that some engineer's idea or was it the architect who decided, "I want to expose a structure?"

WYLLIE: It's more that when the architect has to figure out what kind of bracing system is going to go into the building, and when he and the engineer send the iterations back and forth, it turns out to be the lesser of several evils.

TOMASETTI: I was interested, though, in your saying that it's something you, the engineer, thinks is a good solution, even if it usually is not accepted.

WYLLIE: We sometimes think it makes a good solution. I can think of one building—it never got built, but it was going to be a 10- to 12-story parking garage. You need openness for ventilation, and we had the idea of making the perimeter of cast-in-place concrete with square openings on the diagonal. So you had a series of latticework braces, you had a lot of ventilation, and we thought it would be fairly attractive. I guess we just hit the wrong architect. We got a pretty cold no.

TOMASETTI: It sounds similar to what Skidmore had done in Chicago and in New York. Fazlur Khan came up with the concept initially, a concrete shear wall building with windows that are offset on different floors and form a diagonal pattern of exposed concrete (see Figs. 3.14 and 1.2).

WYLLIE: Like a regular window pattern where certain panels are closed in to create diagonal bracing.

TOMASETTI: Exactly. Why do you think some architects resist ideas about exposing structure?

WYLLIE: I think with some, the idea is much better if it's theirs. A lot of very fine architects have an ego thing. I think all of us engineers could be better communicators and find a way to plant the concept in their minds—then we might be more successful.

TOMASETTI: Getting back to exposed structures in California.

WYLLIE: In thinking back some 20 years, when architectural concrete was not as expensive as it is now, a very common approach on the West Coast was to put your shear wall around the outside, then add a series of piers and spandrels. You get into the concept of whether it's one big wall with some openings or a series of shear walls with coupling beams. There are a lot of buildings built that way. Instead of being true exposed concrete, they may have some kind of veneer or finish or plaster on top to give texture.

You can make a reasonably good argument, with the cost of curtain wall systems today, that concrete shear wall on the perimeter doesn't cost any more than the curtain wall per square foot. And you've provided your closure as well as your structure.

TOMASETTI: That brings us to the question: To what extent does exposing structure allow you to economize?

WYLLIE: When you combine things, like using part of the exterior as the structure and the bracing, you can save some money. But when you pour concrete around the perimeter—with all the innovations in forming now—it can be harder to get the forms out and up to the next floor, to get going as fast. So, from a construction point of view, you box yourself in, and it negates some of the economizing.

TOMASETTI: What are your thoughts on the use of composite concrete and steel construction—on the outside, a concrete tube encasing steel, with interior steel framing and steel core—the type of thing done in Hong Kong on the Bank of China? Is that being done on the West Coast?

WYLLIE: Not too much. I think one of the questions is how well the frame will work in an earthquake. It's really an area where we don't have research. And it doesn't have its niche in the code, as far as cyclic loading and all that.

In terms of steel frame buildings, we've done a lot of mixed structures, some with concrete walls around the core. You make the structural steel part of the composite shear wall system. It's not done so much currently, due to the cost of forming concrete around structural steel.

TOMASETTI: What would your pointers be for an architect who just loved the John Hancock building and wanted to do a very exposed structure in steel? Or, what if you had another case in concrete? What do you think of as important in terms of the California environment?

WYLLIE: I don't know if you necessarily do anything that different in California. The earthquake issue may or may not influence exposed structure. I think there's probably a need to have an added level of attention on any building, exposed or unexposed. And certainly, you've got to address wind design, though it's a lot easier addressed than earthquake forces. In thinking of some of the exposed structures we've done, the architects haven't had as many preconceived notions of what the structures are going to be. You know, architects usually have a pretty good idea of what they want by the time they bring you in.

TOMASETTI: Right, right...

WYLLIE: So many times the structural engineer, when he comes to the first meeting to find out what the building is about, just gets little sketches. The architect doesn't have every last detail down cold, but the concept is pretty well frozen. Here are your column lines, and so forth. Whereas in some of the more successful things, if you get involved early with the architect, all he may know is that it's going to be so many stories, so many square feet per floor, and from the shape of the lot, the order of dimension.

I'm thinking of one structure, about a 30-story building, that was sort of exposed. The architect said, "Before you get too hung up in your concepts, tell me what, ideally, you'd like to do with the structure." There was a lot of flexibility on where the columns could go, so we played around with it and decided to use some diagonal braces; that would certainly keep the cost down. He said wherever I wanted the brace, he'd just make that a wall. He developed a system of precast stucco walls to cover the braces with.

We put the steel package together and tried to get prices. We went out to the steel people, gave them the tonnage, and they looked at the thing and said, "We'll give you a price, but something's wrong; you can't build a building in San Francisco that cheap." So we went back to show them all the sizes and let them take all the quantities. They came up with the identical number.

TOMASETTI: What happened in the end?

WYLLIE: The owner was sued by neighbors and ended up not building the building. It would have been one of the first eccentrically braced frames. We came up with the frame just trying to get the diagonals in and make them work—it was just the system we developed.

TOMASETTI: Was that going to be exposed?

WYLLIE: Not the steel, they were going to cover it with the stucco panel.

TOMASETTI: But it would have been expressed.

WYLLIE: It would have been expressed. It wouldn't have been that different from SOM closing in the windows, but here we were closing in the bays, if you will.

TOMASETTI: I can think of another way in which an exposed or an expressed structure might be different. Take, for example, a reasonably tall building on the East Coast that has a braced-tube frame. If the braced tube is our major wind resistance system, I'm just interested in getting forces against the tube—my diaphragm action. And as long as I have a way of getting the forces out of this major brace on the outside, I'm going to be happy. All these braces are going over several floors at the same time.

I submit that an earthquake design would be a little different, since you have a whole batch of inertial effects to contend with between the sixth floor and where this major brace is going on. The whole building might stay up, but everything in between could blow out.

WYLLIE: We like to make it a little more relaxed. A term we love to use in earthquake resistance is redundancy...there's nothing like redundancy. You don't want to put all your eggs in one basket, in one big brace, if it's close to lamellar tearing problems or locked in welding stresses, or anything like that. You'd much rather have a lot of smaller members. In case one of them fails, all you've got is some welds to repair.

There are some engineers who don't worry about those things much. "Meets code, fine. That's all we have to worry about." We see a lot of that everywhere.

TOMASETTI: That attitude has really got to go. The idea of meeting the code is only a constant reminder that the code is an absolute minimum. Sometimes the owners are cost-conscious, and every once in a while, if code is not enough, we have to take a stand and say, "Just get somebody else, because we're not going to do it that way."

WYLLIE: You asked about other examples of exposed structures I've proposed or designed. Another comes to mind, a wood-frame ski lodge. The architect wanted a flat roof, and in snow country, that's not always the best idea. He wanted a high lobby and rooms on each side, with balconies. We ended up using exposed poles for the columns along the balcony and building up plywood shear walls back in between the rooms. That was no problem for bracing, but then you had the snow load to hold up. So we decided to keep the span down to about 8 feet on the roof, where you'd just be in too much trouble with the snow on the main girders. The lobby was about 25 feet wide, balcony to balcony, so we had the two poles coming up, with the clerestory up at the top. We came up with two diagonal poles that came down a couple of stories, and came in with some steel connections, all exposed.

The architect got all excited about it. The structure was built, and the first winter they got 9 or 10 feet of snow on the roof. It was another whole story of snow up there, and it worked great (Fig. 3.36). We were very conservative in a few of the things we did—on the critical things. You worry about wood, factors like

Figure 3.36 A mix of glued-laminated girders and rougher round poles is visible in this construction photo of a hotel lobby area. The knee braces reduced roof spans to handle snow loads.
Bear Valley Lodge, Calavaras County, Calif; Architect: *Botsai & Overstreet;* Engineer: *H.J. Degenkolb Associates* (Photo: *H.J. Degenkolb Associates*)

crushing, compression perpendicular to grain (Chap. 6).

TOMASETTI: What are some other uses of exposed structure?

WYLLIE: Probably the biggest use on the West Coast right now is in strengthening buildings. There are all sorts of old brick buildings in San Francisco and Los Angeles. To strengthen them you end up putting in exposed steel—steel diagonals—and often you paint them some pretty color. It's by far the most economical way to do that sort of thing. And so out in San Francisco now you see those diagonal braces running behind the windows in those old five-story brick buildings.

The Japanese had a building in Sendai that was damaged a few years ago in an earthquake, and now it's all diagonals on the outside. We've been trying to sell diagonal bracing—it's usually a cheaper system, especially if you have a building that has to remain occupied during the repair work. If you use steel bracing, you do it all from the outside. Drilling holes and epoxying bolts can be a little noisy; you have to control the time when you do that. But you don't have to shut down the building, and it usually saves money, too. Of course, you do have to be a little inventive on the structural shapes you use, so they'll end up looking aesthetically pleasing.

4 *Exposing Concrete*

PROPERTIES OF CONCRETE

Recently a designer was heard to muse, "Wouldn't it be great if there were some material you could just pour into any shape, and it would harden into a structure?" He was brought back to reality with the response, "Yeah. It's called reinforced concrete."

Concrete is such a common part of the urban landscape that sometimes we overlook how versatile a material it can be. It is the workhorse of modern construction by any measure—volume placed, square feet of area built— due to a high strength-to-cost ratio, good durability, ease of placement, and good inherent fire resistance. These characteristics make concrete ideally suited for exposed structures, especially those exposed to the weather. But its key asset for exposed structures is its sculptural property. Cast in forms, it can produce almost any desired shape (Figs. 1.8, 4.1, and 4.2). Exposed faces can be developed into hard surfaces capable of withstanding the wear of traffic or weathering conditions, or elaborately detailed to satisfy an architectural design requirement.

For successful use of exposed concrete, the designer must select structural systems, details, and finishes which take best advantage of its strengths and avoid, minimize, or conceal its weaknesses. Aspects to be considered include mix constituents, workability, strength, deflections, density, corrosion protection, fire resistance, design details, finish, and color. Each aspect will be discussed in brief, with special emphasis on its relationship to exposed structures.

Mix Constituents

Concrete is composed of natural and manufactured products, mixed and placed in individual batches. This can be both an advantage and a disadvantage. By selecting different constituents and by varying their proportions, the resulting concrete can exhibit very different properties. In this way, desired properties can be provided in an economical fashion. However, inadvertent variations in constituents and proportions can result in undesirable

Figure 4.1 This icon of modern architecture takes full advantage of the plasticity of concrete.
TWA Terminal, J. F. Kennedy Airport, New York, N.Y.; Architect: *Eero Saarinen*; Engineer: *Amman and Whitney*. (Photo: *Len Joseph*.)

variations between batches, especially where a close match of color or texture is desired.

Concrete is a mixture of cement, fine aggregate, coarse aggregate, and water. It may also include pozzolans and admixtures. The completed structure also includes steel reinforcement. All these will be discussed in turn.

Cement. Modern portland cement is a hydraulic cement. That is, it hydrates—reacts chemically with water—to gain strength. It is created from limestone and shale, or other calcium carbonate and clay-bearing rocks. They are ground, mixed, heated until particles fuse, and then reground to a fine powder.

Cement can influence concrete color. Iron oxides and manganese oxides occur in most clays and shales, giving most cements a gray or green-gray color. Cements with fewer oxides can be nearly white, with buff, cream, blue, or green undertones. Selection of these special cements may be necessary to achieve particular colors, especially the lighter tones. White cements tend to be low in

Figure 4.2 This office building uses cast-in-place concrete to create drama in a different way. Note the rustication.
Federal Home Loan Bank Board Building, Washington, D.C.; Architect: *Max O. Urbahn Associates*; Engineer: *Lev Zetlin Associates Inc./Thornton-Tomasetti.* (Photo: *Janice Tuchman.*)

---

alkali content, which can be an additional benefit if available aggregate stock is susceptible to alkali-aggregate reactions (see below). Cement color, especially for common gray cement, can vary markedly, depending on the raw material source, manufacturing process, handling, and delivery equipment used. This can affect the uniformity of the finished product. Therefore, the designer should require that all cement for exposed portions of a structure be of the same type, brand, mill, and raw material source. In addition, any cement or concrete samples reviewed for approval should state the type, brand, and source of materials.

Cement formulation can provide variations in strength and chemical action. High-early cement provides faster strength gain, generating additional heat in the process. Rapid strength gain is an advantage to precasters, who want a rapid turnover of forms. Heat generation is helpful when placing concrete in cold climates. Sulfate-resisting cement has improved resistance against chemical attack, as can occur in some soil, groundwater, and industrial locations. It can also prove helpful when exposure to salt water is anticipated, as at coastal bridge abutments and seawalls. Shrinkage-compensating cements expand slightly over time, tending to close any shrinkage cracks which may appear. They have been used successfully in exposed concrete, but should be carefully researched before being specified for any particular application.

For good appearance at exposed faces, the Portland Cement Association (PCA) recommends 335 kg of cement per cubic meter (564 lb per cubic yard or 6 sacks per yard) as a minimum, regardless of strength requirements.

Fine Aggregate. Fine aggregate usually takes the form of naturally occurring sand, but can also be fines from rock-crushing operations or other industrial processes. Fines from crushed lightweight expanded aggregates can be used for an all-lightweight concrete. Fine aggregate improves concrete economy by filling the interstices between coarse aggregate particles, replacing expensive cement. It improves workability by acting as "ball bearings" between coarse aggregate particles, and reduces shrinkage by replacing shrinkage-prone cement-and-water paste with inert material.

While generally a trouble-free constituent of concrete, five aspects of fine aggregates require special attention for exposed structures. Some fine aggregates are chemically reactive, such as quartz, which has been strained or deformed by geologic activity. Alkali-aggregate reactions (AAR) can occur in such cases. The products of these reactions have a greater volume than the original cement and sand, causing swelling and internal microcracking, which leads to deterioration of the entire concrete mass. This becomes a particularly important consideration when the concrete is to be exposed to weather, as AAR is aggravated by the presence of water. In one case, identical concrete members showed no AAR where sheltered from rain, but cracked and crumbled in only a few years where exposed to the weather. Aggregates can be tested for possible AAR using procedures in ASTM C33, C227, and C289. The American Concrete Institute (ACI) recommends that several varieties of tests be performed before assuming that AAR is not a problem.

Second, because fine aggregate normally lies at or near the concrete surface, it strongly affects the color of the concrete, especially when used with white cement. By selecting colored sand, special concrete shades can be achieved with little difficulty. However, note that most natural sands are not white enough for a true white concrete, so crushed limestone or quartzite may be required.

Third, the amount of fine aggregate must be established considering the desired end product. A high proportion of fine aggregate gives good control of color uniformity but may lead to "bug holes," small air pockets on the formed surface (Fig. 4.3). On an exposed-aggregate surface it would also tend to provide a surface which shows more sand-cement matrix and less coarse aggregate.

A fourth aspect of fine aggregate, currently under study, is the thermal incompatibility of concrete constituents (TICC). This can lead to accelerated breakdown of concrete where it is exposed to wide temperature swings. TICC is aggravated when the fine aggregate and coarse aggregate are of different origins.

Fifth, the aggregate should come from one source and supplier throughout the project to minimize variations in color, workability, and strength.

Coarse Aggregate. The type, shape, and size of coarse aggregate used can have a marked effect on the properties of a concrete mix. Hard rocks such as granite and basalt ("trap rock") impart good abrasion resistance and yield concrete with a high elastic modulus (stiffness). Dolomitic limestones, while somewhat softer, are used in concrete with the highest compressive strengths, as the stiffnesses of the rock and of the surrounding cement paste are similar. This avoids differential strain, which can lead to the microcracking and failure at lower stresses observed when using granite and basalt. Sandstones vary widely in hardness, but can be used successfully in concrete in many cases. Slate and shale are avoided in concrete work for two reasons. When crushed, the resulting particles are flat

Figure 4.3 While small "bug holes," or air pockets, are inevitable in concrete construction, some of these are objectionably large. Note pen for scale. (Photo: *Len Joseph*.)

and platey, which makes for weak, hard-to-work, stratified, or laminar concrete. Also, the nature of these rocks, being basically hardened clay, leads to poor performance at the cement paste–aggregate boundary. The aggregate can soften, leading to internal slippage, microcracks, and excessive deflections.

Manufactured coarse aggregates are used in the production of lightweight concrete. Taking a cue from natural pumice and scoria, manufacturers heat carbon-bearing shales, slates, and clays until the stone is plastic and the carbon vaporizes, forming trapped gas bubbles in a spongy mass. Layered mica can be similarly expanded into lightweight vermiculite. Sand-lightweight concrete (only coarse aggregate is "expanded") and all-lightweight concrete (with fine aggregate of crushed expanded aggregate) can have weights of, respectively, 20 and 30 percent less than normal-weight concrete. Lightweight aggregates have been used successfully in many applications. With proper finishing, troweling develops a cement-sand "mortar" over the coarse aggregate, providing a durable surface. One caution is that the porous aggregate will tend to absorb water, so presoaking is recommended and the mix should be designed using such presoaked aggregate to avoid having a concrete which is too dry for placing or finishing. Without presoaking, water absorption problems can be particularly pronounced where concrete placement is by pumping—the pump pressure can force water out of the cement paste and into the aggregate pores.

Aggregate shape can affect mix design, workability, and finished appearance, especially where the aggregate is to be exposed. Naturally rounded aggregate provides the lowest cement demand for a given strength. The rounded shapes pack closely together, leaving less space for cement, sand, and water to fill. The "ball-bearing" effect also provides the most workable, easy-to-place mix.

Naturally rounded aggregate is found in river beds and in glacial deposits such as moraines, drumlins, and kames. Environmental regulations and previous use of these sources are limiting their availability. And, of course, many areas have no such deposits. In those areas where rounded aggregate is unavailable, or the minerals found in such deposits are unsuitable for use in concrete, crushed stone must be used. Here the more cubic the shape, the better the particles will pack. When techniques are used to expose aggregate, rounded aggregate gives a cobblestone effect with deep relief. Crushed-stone aggregate finishes tend to show mainly flat faces, especially if form vibrators are used, giving more of a mosaic effect. The amount of retarder and timing of scrubbing control the amount of aggregate projection. Surface treatments are further discussed at the end of this chapter.

AAR can be a concern with coarse aggregates too. Any aggregate source without a previous "track record" should be extensively tested before acceptance. Here again, exposure to moisture can worsen the effects of AAR.

Another concern with exposed concrete is the porosity of the aggregate. Coarse aggregates which absorb significant amounts of moisture can cause a pockmark effect when exposed to freeze-thaw cycles. Aggregate can be tested for this tendency also, using ASTM C666. Similarly, soft high-calcium stone, including some marbles and limestones, can exhibit poor resistance to surface weathering, especially under acid rain.

Coarse aggregate can strongly influence color, especially when exposed by retarder or sandblasting. Even when not exposed, dark stone can affect light cement. The cement paste will not completely mask aggregate just below the surface, so the resulting tones will be darker than expected. Another item is unintentional color. Some iron-bearing stone is perfectly acceptable for structural concrete use, but tends to create rust spots where exposed to moisture. Guard against this by checking that the source proposed has a successful previous track record. Unexpected color can also come from contaminated stock piles—a little red clay dust can go a long way in a white cement–white limestone mix. Cleanliness at the batch plant and in the transit mixers must be emphasized and monitored.

Water. Portland cement's best friend is water. Cement gains strength by hydration, which requires the presence of water. Curing fresh concrete by covering the surface for several days with a pond of water, wet burlap, or a high-humidity fog provides the most crack-free results. Subsequent soaking will also cause the concrete no harm. However, the water used or applied should not carry deleterious salts or other chemicals. Salt will not seriously damage plain concrete, but it will lead to rusting of embedded steel reinforcement, and the expansive action of the rust will weaken the concrete through splitting along the rebar-concrete interface and microcracking of the concrete itself. Other chemicals, as in sulfides, sulfates, and acid rain, will deteriorate concrete through chemical action.

While hydraulic cements do require some water to enable the strength-creating chemical reactions to occur, concrete mixes typically use more than this amount of water. Chemically speaking, a water-to-cement ratio *w/c* of 0.2 by weight is sufficient. However, the resulting cement paste would be like dry bread dough, impractical for use in the field. Workable mixes have two to three times that much water, that is, a *w/c* of 0.4 to 0.6. The additional water is termed "water of convenience," as it makes the concrete fluid enough to conveniently place in forms and around rebars. Once the concrete has been placed, this excess water serves no function and most of it eventually evaporates. Water of convenience should be minimized, as it leaves microscopic voids once it evaporates. These voids have no strength, so the resulting concrete is weakened. Also, as the excess water evaporates, the cement paste shrinks. If shrinkage occurs while the concrete is "green" (has not yet developed significant strength), widespread visible cracking will occur. Less excess water means less shrinkage and less cracking.

Water can have a surprisingly significant effect on color. The possible presence of high iron content should be checked out early, even in potable (city) water. Rust stains could develop in such a situation. Wetter mixes tend to result in darker finished concrete, so variation in water content from batch to batch should be avoided.

For tall pours, such as walls and parapets, excess water of convenience, or bleedwater, rising to the top of the form will make the top of the pour darker than the bottom, one source of visible pour joints. In theory, a tall wall requiring several truckloads of concrete per pour could benefit from having later (upper) batches mixed drier than early batches, but this could be difficult to control in practice.

For architectural concrete there are other reasons for using a dry mix (w/c less than 0.46). First, a wet, or soupy, mix permits segregation of materials, with coarse aggregate sinking to the bottom and water-cement paste flowing to the top. Obviously, if proportions of stone, sand, cement, and water vary with height, colors and textures will also vary. Second, excess water finds its way out of forms through joints, and this wash-out effect leaves sandy lines behind. Less water (and, of course, tighter form joints—to be discussed later) results in less visible joint lines. Third, soupy mixes are more likely to form bug holes, those small air bubbles which may be visible at the formed surface of any concrete, even well-vibrated concrete (see Fig. 4.3).

After water has been added at the batch plant, mixed in transit, leached from joints, risen in the form, reached the surface, and evaporated, it can still affect concrete color. While curing, the concrete surface should be kept wet, or at least damp, but uneven wetting can lead to uneven shades of concrete. This can be particularly pronounced under plastic blankets, where contact areas can be clearly visible. And, of course, wetting by rain will temporarily darken concrete, with runoff channels potentially showing permanent stains.

Pozzolans. The first documented hydraulic, or water-activated, cement was created by grinding naturally occurring volcanic tuff or ash and was given the name Pozzolana by the Romans after the city Pozzuoli. Today the term pozzolan is applied to non-portland-cement materials which exhibit the characteristics of hydraulic cement to varying degrees.

Fly ash has been used for several decades as a pozzolan. When coal is burned, the noncarbon impurities being carried up the flue are captured by pollution control equipment. This fly ash (as opposed to heavier bottom ash) is mainly silica, which has cooled into microscopic glass beads of somewhat larger size than cement grains. Fly ash can be used as a partial substitute for cement and offers four advantages. First, it is less expensive than the cement it replaces. Second, the "ball-bearing" nature of fly ash results in improved workability at any given w/c ratio. Third, fly ash reacts more slowly than portland cement, an advantage where heat buildup during hydration is a concern, as in hot weather and in massive pours for dams, caissons, and foundation mats. Fourth, the "ball-bearing" action can permit reduction of water while maintaining workability. This results in higher concrete strengths. However, note that a cement–fly ash mix will gain strength more slowly than an all-cement mix. This is no problem where loads are applied slowly, as in foundations and columns, but can be a concern if casting slabs on a fast schedule.

For exposed concrete, fly ash can be a valuable constituent due to its ability to permit reduction of water of convenience. However, its use also requires care. Coming from a chimney, it should be no surprise that fly ash will tend to darken concrete somewhat. The dark color comes from soot (unburned carbon) trapped within or settled on the fly ash particles. For good control of color, and control of concrete properties, fly ash for one project should come from a single, specific power plant. Note that this provides no guarantee of control, as the coal supplying that plant may come from several sources (although this should be less of a concern with the newer "mine-mouth" plants in the Western states). Note also that fly ash can be classified in several groups with different reactivity and different amounts of free carbon (loss on ignition), so the mix supplied to the job should match the mix design submitted for approval in all respects.

Microsilica, or silica fume, is a by-product of silicon manufacture for electronics. These smokestack particles are much finer than those of portland cement, and this fineness gives microsilica some special properties. The high surface-to-volume ratio of the particles means that their chemical reaction with cement is rapid and efficient. They react with free lime in the cement paste to form stronger cementitious products. Concrete with microsilica gains strength rapidly and can reach extraordinarily high levels. The microsilica particles and products of reaction fill pores and gaps between cement particles, blocking the flow of free water through the concrete. Thus concrete with microsilica is more waterproof than conventional concrete, and will better resist the penetration of damaging water-borne salts. However, microsilica is not a cure-all.

Microsilica can have an impact on exposed concrete. It adds significant expense to a concrete mix. The mix can be sticky and may require special finishing techniques. Being a new product, the experience base regarding exposing aggregates, providing special finishes, and providing color is less extensive than for conventional concrete. And the impermeability of microsilica concrete, which is so helpful in blocking water penetration, also blocks the bleedwater from escaping. Bleedwater normally rises to the top surface of a pour and keeps the fresh concrete surface wet for curing. In a microsilica mix, aggressive wet curing measures such as wet burlap and sprinklers must be used to make up for this lack of bleedwater to avoid excessive surface shrinkage cracks. And, once again, mix design and production concrete should use a single source of microsilica to minimize variations.

Admixtures. Other chemicals can be added to the concrete mix to affect its strength, workability, and resistance to damage. Accelerators and retarders affect the rate at which the chemical process of cement hydration occurs. Accelerators are used where rapid strength gain or chemically generated heat is desired. This most commonly occurs in cold weather concreting, where the accelerator offsets the tendency of chemical reactions to slow down at lower temperatures. The "granddaddy" of accelerators, calcium chloride, should never be used with architecturally exposed concrete, or any concrete containing steel and subject to wetting. It can trigger corrosion of reinforcing steel, especially in the presence of water, and it can cause darkening and mottling of the concrete. Several nonchloride accelerators are available instead. Retarders are normally used where limited working time is a concern. In hot-weather concreting, the rapid rate of chemical reaction can cause concrete to "set up" before it can be placed, consolidated, and finished properly. Or where a concrete pour proceeds slowly, as when placing a foundation mat or when placing concrete using small buckets in confined spaces, delayed setup avoids "cold joints," or unanticipated boundaries between just hardened and fresh concrete. The use of retarders to provide an exposed-aggregate finish is discussed later. Retarders should be used with caution, as high dosages can cause cracking, discoloration, and overly delayed set.

Water-reducing (WR) agents and high-range water reducing (HRWR) agents, or superplasticizers, act as lubricants during concrete placement. For a given *w/c* ratio, workability is increased. Alternatively, for a given workability the *w/c* ratio can be reduced and thus the concrete strength increased.

Air-entraining agents are of particular interest for concrete exposed to the weather in cold climates. They improve resistance to freeze-thaw damage. Ordinarily when water trapped in pores of concrete freezes and expands, the force generated tends to locally fracture the concrete. In the next cycle water enters the cracks and freezes, enlarging the cracks. Concrete failure then would proceed from the surface inward. Air-entraining agents create a network of closely spaced microscopic air bubbles throughout the concrete, occupying 4 to 6 percent of its volume. These bubbles serve as relief valves for entrapped freezing water, reducing the internal stresses and protecting the concrete from progressive cracking. While the presence of bubbles slightly reduces concrete strength, air entrainment is so cost-effective that it should be used in any concrete which will be exposed to freezing weather. The bubbles created are invisible to the naked eye, and should not be confused with objectionable bug holes and honeycombing, which can result from poor concrete placement practices such as inadequate vibration. However, high dosage rates of air-entraining agents can cause an increasing frequency of bug holes, so when using these products, aim for the low side of the entrained air target range. Also, as for any admixture, the effect on concrete color should be tested during mix design, using a variety of test panels, and controlled during production by using single sources and standard dosages where possible.

Mineral oxide pigments, although expensive, can be used to produce a variety of exotic colors. Cobalt oxide produces blue, brown oxide of iron produces browns, chromium oxide produces green, red oxide of iron produces red, and black iron oxide produces grays or slate. In its publication, "Color & Texture in Architectural Concrete," the PCA (1980) cautions that the amount of coloring materials added should not exceed 10 percent by weight of the cement, as larger quantities of pigment may excessively reduce the strength of the concrete. Architect Jeremy Wood, senior associate with The Architects Collaborative, Cambridge, Massachusetts, offers a caution regarding the use of pigments and other admixtures to achieve color variations. "You can't control it," he says. "The products are getting better but it's still risky business." Nevertheless, pigments offer the advantage of very wide ranges of color compared to the limited color swings available through colored sands or aggregates.

Reinforcement. While massive blocks of plain concrete may be the hallmark of some civil engineering structures such as dams, virtually all concrete work in buildings contains reinforcement. This is usually in the form of mild-steel deformed reinforcing bars of 280- to 460-MPa (40- to 66-ksi) yield strength, but may also include 60-ksi welded wire mesh, prestressing wires, and multiwire posttensioning strand of 1.9-GPa (270-ksi) strength. Under most conditions concrete and steel work very well together. Their coefficients of thermal expansion are similar, so temperature changes do not induce large internal stresses which would cause the two materials to separate.

Reinforcement of exposed concrete must consider corrosion protection and fire resistance. The highly alkaline environment of the cement paste "passivates" the steel surface, protecting it by inhibiting the process of rusting. Where additional rust protection is required, the rebar can be epoxy-coated or galvanized. Steel is protected in fires by the large thermal mass of concrete combined with the heat absorbed by vaporization of free and adsorbed water in the concrete, and the heat needed to break the chemical bonds of hydrated cement. These topics are discussed later in this chapter and also in Chap. 7. Special considerations for architecturally exposed concrete relate to avoiding unsightly rust stains. Rebar cover is provided by generous use of support chairs and ties, but these must not contribute to the problem themselves. Chairs should be nonmetallic, or should have plastic feet (Fig. 4.4). Lathing tie wire should be soft stainless steel (if using uncoated rebar) or plastic (for epoxy-coated rebar) to avoid rust spots when ties protrude to the concrete surface.

Figure 4.4 When chairs used to support reinforcing bars do not have plastic or noncorroding feet, rust spots like these can occur at soffits exposed to weather. (Photo: *Len Joseph.*)

Workability

The workability of concrete imparts its sculptural property. Workability defines the ease with which ingredients can be mixed and the resulting mix can be handled, transported, and placed with little loss in homogeneity. To contractors, good workability is important because it is integral to the ease of concrete placement, but workability is also a key to minimizing defects in the finished surfaces of exposed structures. For example, a stiff mix may have a low *w/c* ratio and develop great strength, but if it does not flow around the rebar, "honeycombs" and voids may result, or corners may not be filled. Even if the forms become filled, narrow spots in the formwork and closely spaced rebar can act to "strain out" coarse aggregates, leaving some areas with only a sand-cement mortar. This affects strength, shrinkage, and appearance.

For architecturally exposed concrete, workability may involve tradeoffs. For example, as discussed, architectural concrete should have as little water of convenience as possible to minimize bleedwater problems and bug holes. But complex shapes and sharp corners require more workability, which usually means more water. Modern superplasticizers offer some relief from this dilemma, giving workability without added water. Redesign of the most constricted details can make workability less critical. The member shape can be modified, the sequence of pours can be changed, or the formwork can be provided with access hatches to bypass narrow spots. Rebar can also enter this picture, as good workability is needed to get between closely spaced bars. Spacing can be increased by using larger bar sizes, subject to development length and crack control limitations, and congestion at laps can be eased by staggering splices or by using mechanical splices such as Lenton couplers.

Another workability tradeoff is gap-graded coarse aggregate, which is used for better exposed aggregate appearance. This normally produces a "harsh" (less workable) mix, which can be offset by adding fine sand. But high sand content can increase the incidence of bug holes. Here trial batches and trial panels will be required to establish an acceptable compromise.

Workability is usually determined by performing a slump cone test on the fresh concrete just before placement. A tapered cylindrical steel form 300 mm (12 in) high is filled with fresh concrete in a prescribed manner, and then the form is lifted off and the free-standing cone of concrete spreads and "slumps" down. The decrease in height is measured as inches of slump. A slump of 75 to 100 mm

---

Chapter Four

(3 to 4 in) is relatively stiff but workable if vibrated, 127 to 178 mm (5 to 7 in) is very workable, and greater than 7 in is "soupy" enough that segregation and excess bleedwater may cause problems.

While slump is commonly used in the United States as the measure of workability, it may not be sufficient for the special needs of exposed concrete work. Tests with full-size mockups of critical areas should be performed to confirm acceptable form-filling ability and appearance before settling on a particular design mix.

Strength

Here again, concrete provides the opportunity to trade off various desired characteristics, and the adage "you can't have too much of a good thing" does not apply. The strongest influence on concrete strength is the water-to-cement ratio w/c. Low w/c, 0.40 or less by weight, gives relatively stiff, strong concretes, while high w/c mixes, 0.50 or more, are very workable but of lower strength. Note that a low w/c can be achieved by reducing the water in the mix. This can make an unworkably stiff mix, or may require the additional cost of an HRWR (superplasticizer) admixture. Alternatively, the water level can be maintained, but cement can be added (replacing sand). This adds cost also, and the increased volume of cement paste increases the potential for concrete shrinkage. Relative costs of the two approaches vary with producers and locations. Influence on appearance should also be evaluated.

Concrete strength does not eliminate other potential concerns. Concrete can be strong but still subject to freeze-thaw damage, so air entrainment is still recommended for cold-climate exposure. In fact, increased cement paste may require larger doses of the air-entraining admixture. Concrete can be strong but still subject to shrinkage cracking. In fact, the strongest concretes, with microsilica, are especially susceptible to plastic (initial) shrinkage cracking. Therefore careful curing procedures should still be followed. Concrete can be strong but still permeable if strength is gained by adding cement rather than reducing mixing water. And, finally, a concrete design mix can be strong, but if the mix is unworkable and extra water is surreptitiously added at the job, both strength and quality control are lost. Thus all aspects of concrete design must be considered before specifying a particular strength.

Deflections and Volumetric Changes

Designing for deflections and volumetric changes is an important aspect of any concrete structure. For exposed structures a few additional items must be considered. Background for these items is provided here, while additional recommendations on layout are given in Chaps. 5 and 7.

Elastic Deformation. Any structure, or any object, deforms when loaded. Mount Everest is slightly shorter when mountaineers stand on its peak (not that this diminishes their achievement!). Elastic deformation, whether as column shortening or as beam bending, occurs immediately when a load is first applied. Some elastic deformation is common to all construction materials. The deformation can be calculated quite accurately for steel and reasonably well for concrete.

A difference between concrete and steel is that concrete itself is weak in tension. For this reason steel reinforcing bars are used in concrete tension zones such as beam faces. The concrete at tension faces develops cracks as the rebar steel stretches. For exposed structures, these areas require special attention to ensure that cracks are well distributed and small, not widely spaced and large. Large cracks would be visually objectionable and likely to admit water to rust the rebar. Crack control guidance is provided by the ACI 318-89 code through a factor Z, which encourages the use of smaller rebar sizes and closer rebar spacing for exposed structures.

Creep. A feature of concrete (and, to a lesser degree, wood) is creep deformation. Creep is strain (length change) which causes columns to shorten or beams to bend additional amounts, beyond elastic deformation, over time. The rate of additional deformation tapers off over time (Fig. 4.5). Concrete gets its strength from crystalline fibers growing out from cement grains, interlocking grain to grain and bridging around voids (as from water of convenience). Creep is reduced if concrete is older when loaded, as more interlocking has already occurred. Creep is also reduced if concrete is kept moist, as water-filled chambers can bear some load while voids cannot.

Exposed structure relates to creep in two ways. First, since the rate of strength gain is related to temperature, concrete exposed to high temperature will act "older" and creep less than concrete of the same age exposed to low temperatures. Second, the relative humidity of the concrete will eventually reflect the relative humidity of the environment. Concrete exposed to the rain and high humidity of the Southeast will creep less than that exposed to the dry Southwest.

Creep can provide a hidden advantage. In a system with concrete columns and concrete slabs or beams, differential column creep will induce moments in floors and beams. However, as the differential

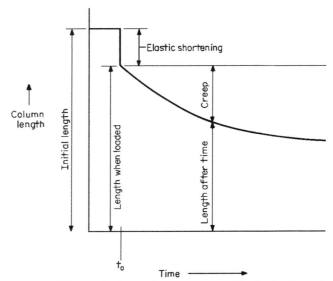

Figure 4.5 This graph of concrete creep shows an idealized pattern of column shortening (t_0—time when load is initially applied).

deflections occur slowly, beam creep will serve to let the beams "relax," and the moments actually developed may be only half that for the case without beam creep. If differential column deflection is caused by temperature changes, creep may reduce induced beam movements due to seasonal temperature swings. Creep is too slow a process to respond to daily temperature swings.

Shrinkage. Another feature of concrete is shrinkage, which occurs as the concrete goes from its initial saturated state to its long-term drier state. When saturated, some water reacts with cement grains chemically to form new products, some is adsorbed or attracted to the surfaces of cement and aggregate particles, and some remains in a free state. This free water of convenience leaves microscopic pores behind as it evaporates, and surface tension across these pores causes them to shrink, compressing and shrinking the entire mass of concrete.

The amount of shrinkage is directly related to the difference between the original amount of free water and the equilibrium, long-term free water in the concrete. Thus more initial water can potentially lead to more shrinkage, and humid ambient conditions mean less water loss and less shrinkage.

If more strength and stiffness is developed in the concrete before water loss begins, the concrete will shorten less as it resists the tendency to shrink induced by surface tension in pores. In addition, as water loss occurs from the outside in, outer concrete tries to shrink while interior concrete resists it, leading to outer tension and outer shrinkage cracks. Stronger concrete has more tensile strength and can better resist outer tension stress, reducing cracking. Both aspects of shrinkage are improved by delaying water loss as long as possible through good curing practices.

Since water loss proceeds from outside in, the size of a member (distance to center) and the shape of a member (surface-to-volume ratio) affect the rate at which shrinkage occurs. Members which are thin, such as floor slabs, or have a high surface-to-volume ratio, such as small columns, experience most of their shrinkage within weeks or months after curing ends. The massive columns of high-rise buildings, in contrast, may shrink more slowly, over a period of many months to years.

Thermal Movements. Concrete has a coefficient of thermal expansion only slightly smaller than that of steel. For concrete members exposed to the weather, seasonal temperature changes will cause significant movements and internal stresses. The relationship between temperature changes and overall building movement discussed in Chap. 5 is generally applicable to exposed concrete, with three notable differences.

First, concrete members tend to be significantly more massive than steel members of equivalent strength. Therefore, a given amount of heat will change the temperature of the concrete member less than that of the steel. Second, the rate of conduction of heat is slower for concrete than for steel. Third, exposed concrete structure in buildings often takes the form of exposed faces of walls, beams, and columns. Thus one side of the member experiences temperature swings while the other side is moderated by the climate-controlled interior.

The thermal mass of concrete tends to dampen out daily temperature swings within the concrete. The lower rate of conduction tends to reduce the influence of outdoor temperature changes on interior heating and cooling demands, and where concrete is exposed to both interior and exterior temperatures, the overall thermal movement of the member will be related to the average temperature, not the exterior temperature alone. As a result of these three effects, concrete structures exhibit thermal movements which reflect seasonal changes more than daily changes, and movements will be of smaller magnitude than for steel alone.

The same properties which tend to reduce thermal movements for a given temperature swing can also act to cause problems when different parts of a member are at different temperatures. A classic situation at open parking decks is "sun cambering," when the sun heats the roof deck upper surface while the lower surface remains cool. As discussed in Chap. 7, the differential temperature and differential strain induce the slab to "hump," or become cambered at midspan. For slabs and beams with fixed ends, this humping creates potentially significant forces in the beam ends and connected columns (see Fig. 7.1). Another common situation is at projecting slabs, such as balconies. The outer portion of slab tries to move during temperature changes, but the interior portion is at a constant temperature. The slab shear forces generated between the two portions must be designed for. This can be particularly significant for precast structures, where connections between portions of slab must be individually engineered and constructed.

Chemical Changes. Concrete can expand significantly when the cement reacts with the aggregate. This detrimental process, termed alkali-aggregate reaction (AAR), has been discussed earlier in this chapter.

Density

Concrete density is controlled by the type of aggregates used. Structural concrete can vary from a density of 1.44 (90 lb/ft^3 [pcf]) when using all-lightweight aggregate, to 1.84 (115 pcf) using sand-lightweight aggregate, to 2.32 (145 pcf) using normal-weight aggregate (stone). Lower density, as low as 0.32 (20 pcf), can be achieved in foamed concretes with gas bubble "aggregate." The bubbles can be air, "whipped in" and stabilized with plastic-related chemicals, or hydrogen, resulting from aluminum powder reacting with the cement. These very light concretes are generally used for insulating or floor-leveling purposes and are not considered structural. Higher than normal density can be achieved by using heavier aggregates, such as scrap steel punchings, but this is generally done only for improved shielding against radioactivity or for counterweight purposes.

For exposed structures, normal-weight concrete is the most frequent choice, but lightweight concrete could prove useful when:

- The desired finish will provide a mortar layer over the aggregate.
- The lower weight is advantageous (as where seismic design or a costly foundation system is required).
- Improved thermal resistance is advantageous.
- The advantages offset the cost premium involved.

Corrosion Protection

For most exposed structures, the alkaline environment provided by portland cement paste is all the protection needed for steel reinforcing bars. In this environment the oxidation process, which leads to rust, cannot proceed. However, the rebar must be well encased in concrete, for several reasons. Carbon dioxide reacts with concrete, reducing the protective alkalinity, but this process proceeds slowly only from the surface inward. Concrete does crack, admitting corrosive oxygen and water, but in properly designed members these cracks decrease in width away from the surface. And oxygen and water, which permeate through concrete, are limited and slowed as they work their way down from the surface. For all three reasons it is particularly important that the minimum concrete cover requirements of ACI 318-89 be carefully followed.

In aggressively corrosive environments, such as chemical treatment tanks and areas exposed to chlorides through deicing salts or ocean spray, more elaborate protective measures should be taken. Treatment tanks are beyond the scope of this book, but chloride protection is discussed in Chap. 7.

Fire Resistance

Concrete offers a three-pronged approach to fire protection. First, the large thermal mass of concrete means that the temperature within the concrete will rise much more slowly than the temperature of the fire. Second, free water held in concrete pores and adsorbed water held by attraction as a film around cement particles are driven off when the concrete temperature exceeds the boiling point. The heat of vaporization of water is high, and by this process a large amount of heat can be received by the concrete and carried away without an increase in concrete temperature. At this point the concrete can still function as intended. The third prong is sacrificial. If necessary, additional heat can be chemically absorbed by breaking down the hydrated cement into its constituents and driving away the hydration water. At this point, of course, the concrete surface will begin to crumble into sand and gravel, but as the process proceeds from the surface inward, enough interior concrete may remain to carry the load.

Reinforcing steel is a significant contributor to strength in concrete columns, and is essential for the strength of concrete beams. As mild steel starts to lose strength at 315°C (600°F), limiting steel temperature is an important part of fire resistance. Here again, as with corrosion, the best protection is adequate concrete cover.

Concrete provides protection so effectively that it is difficult to conceive of a situation where additional protection would be provided by external fireproofing rather than by simply increasing the concrete cover.

Design Details

The secret to effective, economical concrete construction is in developing details which accommodate pour sequences and schedules with minimum custom formwork, special splices, and so forth. The location of construction joints, use of blockouts, type of beam pockets, use of dowels, and sequence of pours are normally worked out by the concrete contractor. However, for exposed structural concrete the architect and engineer should think through the range of possible details which may be desired by the contractor, and establish which ones will or will not be acceptable for appearance reasons. The classic example is an upset spandrel beam at a slab edge (Fig. 4.6). Normally most contractors would plan to cast the beam bottom with the slab, then come back for the upset portion of the beam. However, this would result in an unsightly "cold joint" between the two pours (Fig. 4.6a). A brute-force approach would be

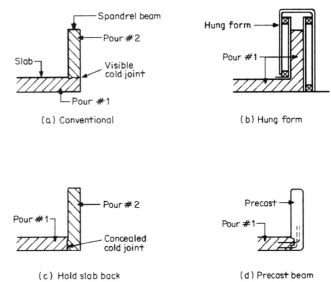

Figure 4.6 Several approaches can be used to construct upset spandrels. (*a*) Conventional with two pours. (*b*) Hung form with one pour. (*c*) Two pours, holding slab edge back. (*d*) Pouring slab against precast spandrel.

to require the construction of beam and slab together. This would certainly eliminate the joint, but would require very expensive and awkward "hung" forms for the beam (Fig.4.6b). Two more economical approaches could be considered. The slab could be poured, but held back from the beam and provided with construction joint dowels, and then the beam could be cast in its entirety (Fig. 4.6c). Alternatively, the beam could be cast first and then the slab cast against it, eliminating the cost of the slab edge form. A variation of this last approach was used at the Westin Hotel in Boston, where the contractor chose to place precast spandrel beams first, with keys and dowels projecting into the building, and then cast the floor slab against them (Fig. 4.6d).

Joints can also be used to regulate hairline cracks and pour lines in cast-in-place concrete. After each truckload of concrete is poured into the formwork, there is a delay. The concrete begins to set slightly before the next batch is delivered, creating a pour line. Horizontal joints can disguise the tendency of a concrete wall to "look like a marble cake." Since concrete is generally placed in lifts of about 2.4 m (8 ft), it is advisable to have a corresponding horizontal joint for each lift. The joint also conceals the ragged break line preferred for mechanical bonding as the contractor goes from one day's pour to the next. Jeremy Wood suggests that a little color variation from one truck load to the next also helps the appearance.

Rustication joints are an architectural feature, but can also serve as a weather-control feature. The joint is a recess in the concrete, which, if properly detailed with the right finish, causes any water that hits the face of the concrete to go into a channel, instead of breaking up irregularly and leaving its mark on the face of the concrete. The exposed concrete and granite facade of the Federal Home Loan Bank Board Building in Washington, D.C., for example, looks good at 16 years of age, partly due to the careful use of rustication joints (Figs. 4.7 and 4.8). The "drip," or groove, provided at the outer edge of exposed soffits, is a specialized weather-control joint (Fig. 4.9).

Because so many factors influence concrete appearance, getting representative samples requires care and good planning. For example, the common practice of producing samples that are 0.6 by 0.6 m by 75 mm (2 by 2 ft square, 3 in thick) or less can lead to misunderstandings. Because hydration rates in the concrete can change the color, Wood suggests that initial samples be a minimum of 0.6 by 0.6 m by 150 mm (2 by 2 ft square, 6 in thick) for precast and 300 mm (1 ft) thick for cast-in-place applications. Prior to construction, cast panels up to 1.2 by 1.8 m (4 by 6 ft), using the same orientation as

Figure 4.7 This 16-year-old building still looks good, due in part to careful detailing and construction. While bug holes are present, they are not visually objectionable. **Federal Home Loan Bank Board Building, Washington, D.C.**; Architect: *Max O. Urbahn Associates*; Engineer: *Lev Zetlin Associates Inc./Thornton-Tomasetti.* (Photo: *Janice Tuchman.*)

Figure 4.8 This detail shows good practice—a drip groove limits staining to the outer edge, bug holes are visible but not objectionable, and form boards are held in plane and well matched at the corner.
Federal Home Loan Bank Board Building, Washington, D.C.; Architect: *Max O. Urbahn Associates*; Engineer: *Lev Zetlin Associates Inc./Thornton-Tomasetti.* (Photo: *Janice Tuchman.*)

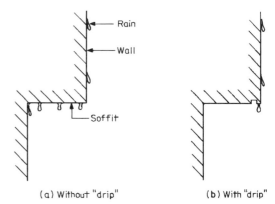

Figure 4.9 Value of a soffit drip. (*a*) At wall section without drip, water can run along soffits and drop at random due to surface tension. (*b*) At wall section with drip, a continuous drip groove (approximately 20 by 20 mm, or 3/4 by 3/4 in) stops water travel across soffit.

proposed for production. For example, a precast panel's appearance requires that all the elements in the mix come to the face of the panel, which is normally cast face down. Similarly, samples for vertical cast-in-place components should be cast vertically. Note that the same concrete mix used for a section of precast will often differ in color, texture, and shading when poured into a column form. This can be a particular concern when precast panels must match cast-in-place exposed concrete structural members, but with care the color can be controlled.

As a practical matter, experienced architects request that contractors start exposed work at the back of the building where mistakes can be better tolerated. In this way the contractor can get experience and the architect can see the nature and range of the as-built concrete. The best architects recognize that this material can exhibit a range of defects, and they design around the defects.

Designers should also recognize that to build a concrete building, you first must build a wooden building—the formwork. Consider the carpentry and the materials—sheets of plywood, generally. Regardless of the building module, in the United States the formwork module will reflect 4- by 8-ft sheets of plywood, and the designer must address that fact. Metal and fiberglass forms similarly come in standard sizes, and there will be seams where panels meet. Economically successful designs capitalize on formwork modules when organizing the expression of the building.

Some types of texture can be provided at concrete faces by formwork. Lumber can be smooth, rough-sawed, or sandblasted to transfer textures to the concrete. Rope can be mounted on plywood forms to produce a ropelike finish. Ribs can be nailed to forms. And plastic form liners, nailed, stapled, or glued to wood forms or glued to steel forms, can produce a number of different textures.

Because fluid concrete exerts a high pressure on the formwork, forms must be restrained, either by exterior framing or by form ties crossing the pour. Form ties are commonly used. Tie marks are left when the forms are removed, so the patterns should be organized to achieve the desired appearance in the structure (see Figs. 3.21, 3.27, and 6.1). Tie-hole patterns can be shown and dimensioned on the architectural elevations and details, reviewed on the formwork shop drawings, and insisted upon in the field. Of course, the project specifications must clearly call out the requirement to meet spacings established by the architect.

Designing and detailing formwork to get the desired results requires careful attention and insistence on proper performance. For example, clean, sharp corners are surprisingly difficult to

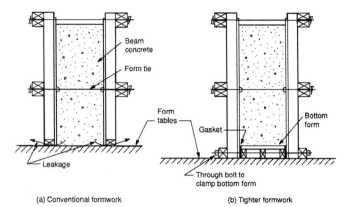

Beam
concrete

Form tie

Form
tables

Gasket

Leakage

(a) Conventional formwork

Bottom
form

Through bolt to
clamp bottom form

(b) Tighter formwork

Figure 4.10 Reducing formwork leakage. (*a*) Conventional formwork sits on form tables. The form-to-table joint is difficult to seal. (*b*) If the beam bottom has its own form, gasketed against the side forms, the joint can be clamped tight.

achieve in cast-in-place concrete. This is because bleedwater runs out through joints, carrying cement paste, which hardens in the joint to form a fin. During stripping the fins break off, leaving rough edges. Chamfer strips are commonly used at corners to reduce fragile 90° edges to more robust 135° turns, but leaks at the chamfer can result in a fragile fin nonetheless. Similar broken edges commonly occur at bottoms of beam forms, where side forms, which simply sit on the form table, leak. To stop leaks, forms must have tight joints. Gaskets, double bottoms, and clamping bolts can be used (Fig. 4.10). The contractor may also have other effective methods. Where forming is critical, full-scale mockups are recommended. Any requirements which cost more than common forming must be specified and explained to the contractor to avoid disputes and delays.

To avoid tie holes, joint patterns, and field quality-control issues, precast concrete can be used. When choosing a precaster, note that different firms have different-size beds. Depth may be just as critical as overall size to allow half rounds, channels, or any other complexity in the shape. Where feasible, ask precasters to cast samples to the architect's design as a condition of bidding. Mockup samples should be at least 1.2 by 1.8 m (4 by 6 ft) and show typical joints, finishes, and end conditions. The design architect should personally meet with the precaster, the detailers, and the formwork supervisors to actively develop the detailing of the formwork as a team. Formwork is a craft, so listen to the craftsmen. Frequently their suggestions will improve on an architect's first conception. Full bay panels should be specified for final approval. Specify two panels, one to stay at the precaster's yard as a control and one to go to the site.

Finish

Coarse aggregate exerts its greatest influence on concrete appearance where the aggregate is intentionally exposed. Otherwise, when cast against forms or troweled on a horizontal surface, cement paste tends to cover the aggregate. The basic methods of exposing aggregate are surface retardation, water washing, brushing, acid etching, abrasive blasting (sandblasting), bush-hammering, tooling, grinding, polishing, and water blasting. These procedures expose the aggregate after casting. In the *retardation* approach, the formwork is sprayed with a retarding agent just before placing concrete. Then, after the formwork is stripped, the face is hosed down and scrubbed, removing the outermost fraction of an inch of cement paste which was kept from hardening. When applied with rounded aggregate, a cobblestone appearance is the result. Most commonly this treatment is used on precast panels with retarder on the bottom form, since treating vertical forms could result in retarder running off to the wrong areas and weakening concrete joints. In a special approach for retardation on vertical surfaces, water-insoluble retarder is sprayed on formboard lying flat and allowed to dry. Then the form is set vertically and used for casting without worry of runoff. The other deep-relief exposure techniques offer their own advantages and disadvantages. *Water washing* or brushing removes cement paste on green concrete, before it has set. As such, the effect is similar to retardation but without much latitude in timing. *Acid etching* removes hardened cement paste chemically. It can also give deep relief, but may temporarily bleach the aggregate. While timing is not critical, the process is laborious. *Sandblasting* can produce deep relief, but maintaining uniformity requires skill, and blasting may expose previously hidden bug holes. The aggregate will be given a matte finish in this process.

For less rounded aggregate, such as crushed stone, low-relief exposure methods come into their own. *Bush-hammering* requires a close match of paste and aggregate hardness and the concrete must reach a good proportion of its design strength. (Otherwise the hammer will mash the concrete and not spall it.) It is not recommended for rounded natural gravels since they shatter and loosen in their sockets. The bush-hammered look is one of fractured faces through both paste and aggregate. *Ground* and polished faces give a very sleek finished look, but require special tools. *Water blasting* is wet sandblasting. It avoids dust problems and simplifies cleanup.

Aside from the various exposure methods used, aggregate appearance can be affected by actions before and during placement. For example, featured aggregate can be hand-laid on a sand bed prior to casting concrete, in the "face down" approach. This requires time and skill for a pleasing final product, similar to field-stone masonry. If casting "face up," featured aggregate is "seeded" onto freshly cast concrete and tamped into place. Here a larger proportion of the sand-cement matrix will be visible. A third approach is preplaced aggregate concrete, in which aggregate is poured into formwork, then a sand-cement-water grout is injected between the stones to fill voids and create concrete. The advantage here is the assurance that aggregate will be lying at the surface in a dense, uniform pattern and can easily be exposed.

The designer can also control the nature of the exposed aggregate in another way. For a more uniform look, the aggregate can be "gap graded," meaning that it is run through sieves to provide, for example, only stones between 20 and 40 mm in diameter. Adding small stones to "nestle" in between large stones would reduce cement demand and make a more workable mix, but this is not done because of the danger that sizes will sort themselves out in the storage bin and cause variable results.

For any type of finish, use of an architectural concrete surface must address the issue of how many imperfections and how much variation of color and texture are acceptable. PCA has suggested that an architecturally acceptable concrete surface be defined as a surface with minimal color and texture variation and minimal surface defects when viewed at 6 m (20 ft). PCA says small blowholes are considered common to concrete and are generally acceptable. Attempts to patch areas of excessive blowholes may fill the holes but produce a patch that looks worse overall. This type of defect is less noticeable in rough textured and exposed aggregate concrete.

Color

As discussed in connection with mix constituents, concrete color can be controlled by choice of cement, fine aggregate, coarse aggregate, tinting admixtures, or a combination of these. Investigating and applying local materials can be a significant cost savings in exposed concrete structure and cladding. A practical designer enters an area with an open mind as to color and texture. Very often the surroundings suggest appropriate matching materials, and if the older buildings used native stone, there is a strong likelihood that matching aggregate may be available locally at reasonable cost.

On the other hand, Wood offers the example of an owner in Cleveland who wanted a white marble or white-metal-clad building. "We went through a costing exercise with Turner Construction of Cleveland and came to the conclusion that precast was the way to go. We paid for white cement and for trucking in white marble chip from Wyoming for the exposed white-on-white aggregate. We paid dearly for them, but the cost was still below the other alternatives and it provided the look the owner wanted."

The other problem to consider, however, is that white buildings don't stay clean for long, especially in industrial cities such as Cleveland. Concrete is going to weather, and it's going to be wet. "Any time you make samples of exposed concrete, you have to hose them down and see what the building will look like wet," Wood adds.

Clear coatings or sealers can be used to prevent the absorption of moisture into the concrete surface. PCA recommends coatings that consist of methyl methacrylate acrylic resin or ethyl acrylate copolymer. Sealers are more fully discussed in Chap. 7. During construction, sealers protect precast units from dirt, rust stains, and grease from equipment used to handle the units. On exposed aggregate surfaces, clear coatings prevent soiling or discoloration by air pollution, ease cleaning, brighten the color of the aggregate, and make the surface water-repellent. PCA cautions that coatings should be selected carefully because some coating materials may cause permanent discoloration. It advises that any coating used should be guaranteed by the manufacturer not to stain, soil, darken, or discolor an exposed aggregate finish.

Wood adds that there is also a maintenance factor involved. "It can have wonderful effects on various aspects of the building, but it's a process that has to be repeated." He cautions that some sealers can also change color with exposure to sunlight. "If you are going to do it, you should see some installations of the product."

Paint can also be applied to exposed concrete to hide a host of blemishes, defects, and variations in color and texture. Purists may not consider this literally "architectural" concrete, but for the purposes of this book it will be considered, because the structure does remain susceptible to temperature and environmental conditions. At Onterie Center the painted concrete frame and cross bracing are visible and exposed to weather (see Fig. 3.14).

An aspect of color rarely considered is patching. Local repairs are inevitable, and patches should match the original work as well as possible. PCA cautions that patching requires skill and close attention to matching the surrounding area. A small amount of white cement—25 percent on the average—normally must be added to the patching mix to duplicate the original color because the lower water-to-cement ratio of the patch causes it to dry darker than the original concrete. If the original concrete matrix is white, on the other hand, a small amount of gray cement may have to be added to give the patching mix the slight gray tint imparted to the white concrete by drum wear during mixing. Trial mixes are essential to determine exact quantities.

CASE HISTORIES

Exposed structural elements and exposed cladding panels have much in common, particularly in terms of creating color and texture. The 120,000 m² (1.3 million ft²) Liberty Center office and hotel complex in Pittsburgh, for example, is clad in architectural precast panels throughout, where a Canadian aggregate was used to produce a pinkish color. The precast was carefully detailed by The Architects Collaborative (TAC), Cambridge, Massachusetts, to complement the historic masonry and terra cotta in the area. The exterior bearing walls are finished with exposed aggregate— Ottawa red granite of 10 to 16 mm—that was sandblasted smooth for the finish. Floors and deck slabs are precast, prestressed concrete members, typically 100 or 200 mm (4 or 8 in) thick. Both the faces of the hollow core walls and the underside of the deck slabs are precast with smooth dense surfaces, and in stairwells these surfaces are left exposed.

A similar case was the Westin Hotel at Copley Place in Boston, which used a special concrete mix on the exposed cast-in-place concrete structural elements to match the color of precast-panel cladding. The precast cladding was chosen for its low cost, erection speed, and aesthetic compatibility with the historic 19th century masonry architecture at Copley Square. Here TAC selected the detail, color, and scale of the precast panels to use them as "cast stone." Glens Falls type III gray cement had been chosen for cost and construction reasons, but to "warm" the color, various sand and aggregates were explored in test panels. Color combinations of smooth and exposed aggregate finishes were developed which would relate to neighboring buildings. A Canadian red granite exposed aggregate was chosen in combination with a Manchester, Connecticut, red sand to produce a pinkish tone. Figure 4.11 shows

Figure 4.11 These two precast panels exhibit three different finishes with a high standard of quality. Note pen for scale.
Westin Hotel at Copley Place, Boston, Mass.: Architect: *The Architects Collaborative;* Engineers: *Lev Zetlin Associates Inc./Thornton-Tomasetti.* (Photo: *Len Joseph)*

a closeup of panels exhibiting three different surface finishes.

The "start-at-the-back" approach to construction quality control was applied at this hotel. On two spiral ramps on the two-level parking garage below the building, Wood used the same detail that he ultimately wanted for the cast-in-place bottom of the building. "It gave the contractor a way of practicing with the formwork and the color in areas where we did not care as much about the appearance." Wood says he learned the technique from Paul Rudolph, who "always had areas that you never see—basements or machine rooms—where the same formwork would be used as for the outside finish."

CONCLUSION

Whether cast in place, precast, or used as cast-stone detail elements, concrete is a logical choice of building material to expose. Composed of inexpensive natural materials, resistant to weathering and fire, ideally suited to fill any mold and take any shape, concrete can offer designers economy, freedom of form, a wide variety of finishes, and a wide range of properties.

Paradoxically, the ease with which this apparently simple, "low-tech" material can be altered to suit particular requirements of strength, color, finish, or construction conditions means that its production and placement must be carefully planned and controlled to get the desired results.

5 *Exposing Steel*

PROPERTIES OF STEEL

Basic properties of steel largely determine its behavior when exposed to the environment. The most important properties include elastic modulus, yield point, ultimate strength, ductility, toughness, fatigue, fire resistance, corrosion resistance, and coefficient of thermal expansion.

A typical stress-strain diagram (Fig. 5.1) illustrates four properties of structural steel. First, the steep slope rising from left to right shows that steel behaves elastically up to its yield point. The slope determines the elastic modulus. Second, the horizontal line indicates a plateau of ductility where steel yields plastically at constant load. Third, steel develops additional resistance from strain hardening and can take additional load until it reaches its ultimate strength. Fourth, the total strain which a particular steel will sus-

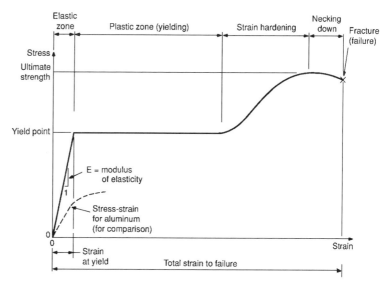

Figure 5.1 A typical stress-strain diagram for structural steel in tension exhibits several distinct zones of behavior depending on strain. Stress=force/original area; strain=change in length/original length.

tain before failure is a measure of its ductility, or ability to be deformed without fracturing.

Elastic Modulus

Steel, a high-strength material, has a nearly constant modulus of elasticity. The modulus of elasticity is the relation between stress (change in load per unit area, expressed in pascals, or newtons per square meter, or pounds per square inch) and strain (the resulting change in length or stretch per unit length). Different types of steels used in construction have yield strengths varying from 250 to 350 MPa (36,000 to 60,000 lb/in^2), but the modulus of elasticity of all the types is the same, 200 GPa (30 million lb/in^2).

The elastic modulus indicates the "springiness" of a material. Being 3 times stiffer than an equal area of aluminum, 8 to 10 times stiffer than concrete, and 20 to 30 times stiffer than wood, steel is an excellent choice where spans are long and deflection control is a concern. In addition the straight nature of the stress-strain curve up to the yield point means that steel acts as a nearly perfect "spring," so deflections can be accurately predicted, whether load is applied one time or in a repetitive fashion.

Yield Point

The yield point defines the maximum stress that a material can resist while still behaving elastically (as a spring). Since additional load causes the material to stretch uncontrollably, somewhat like pulling taffy, the yield point is the upper limit of useful strength for most applications.

Ultimate Strength

Ultimate strength is higher than the yield point. This provides an additional margin of safety against failure. The ductility, or large strain involved before reaching ultimate failure, also provides warning of distress and forgiveness of local overstresses.

Ductility

The ductility of steel can be used to advantage in several ways. First, in the event of unanticipated overloading, ductility will permit significant yielding and deflection to occur before failure. This provides a warning of a potential problem and, if necessary, "running time" for occupants to reach safety.

Second, as when pulling taffy, yielding along the plateau of ductility absorbs large amounts of energy. This is explicitly considered in seismic design as a way to economically provide building safety in earthquake-prone areas. Structures designed to resist seismic forces without yielding are rather expensive, and this approach is used only for critical applications such as nuclear power plants. By intentionally designing for certain selected members and joints to yield and absorb energy in large earthquakes, structural frames for most types of buildings can be provided more economically.

Third, ductility provides for a redistribution of forces and stresses among members and connections. The satisfactory behavior of steel at large bolted connections, where an "exact" analysis would show some bolts overloaded and others lightly loaded, depends on the fact that yielding at some bolts and bolt holes permits others to pick up their fair share. In members themselves, exact analysis would indicate excessively high local stresses at corners of copes and rectangular web openings, but ductility permits these peak stresses to "wash out." Note that such stress concentrations require special treatment when working with less ductile materials such as glass-fiber reinforced plastics (GFRP, or fiberglass), as discussed in Chap. 6. Overall frame action can also benefit from redistribution through ductility. For example, many structures have been designed and built for building code wind forces that were based on a number of simplifications and assumptions. Beginning in the 1960s, and especially in the 1970s and 1980s, a wealth of data from boundary-layer wind tunnel studies for new buildings has shown that code forces are often too low. However, structures built earlier have performed well due in part to the reserve strength and redistribution ability provided by ductility.

Fourth, ductility permits the creation of intentional or unintended "hinges" to relieve load. This explains the satisfactory behavior of many exposed structures which did not explicitly consider or accommodate thermal movement. Pairs of internally generated opposing forces, as in frame legs of a bent, can be limited and relieved by ductility without impairing resistance against externally applied load such as wind.

Elastic modulus, yield point, and ultimate strength are not significantly affected by the normal temperature ranges to which buildings are subjected—about –35 to 70°C (–30 to 160°F) on the surface of the material in an ambient environment of –35 to 50°C (–30 to 120°F) (temperate climates).

Toughness

One property of steel—toughness—is temperature-sensitive. An awareness of the difference between ductility and toughness and of how toughness is affected by temperature is crucial in producing a successful exposed steel design.

Toughness is the ability of steel to absorb energy, particularly in the presence of a sharp notch. In fact, it is quantified by the amount of energy (from a falling or swinging weight) a bar of steel can absorb in the presence of a precisely shaped notch, the classic Charpy V-notch test. Tough materials resist the formation and propagation of cracks at stress concentrations and around imperfections. Materials with low toughness have less resistance and are often referred to as brittle, as fractures caused by lack of toughness are called brittle fractures. Since normal details and methods of fabrication create numerous microscopic cracks and stress concentrations, toughness is needed to avoid fractures in use. Most types of steel have adequate toughness for routine circumstances at room temperature. However, steel exhibits a decrease in toughness at lower temperatures. Each type of steel shows a rapid drop in toughness at a transition temperature unique to it.

As Figs. 5.2 and 5.3 show, the "upper-shelf" impact energy, the steepness of the drop, and the transition temperature of the drop are affected by the alloying elements such as carbon (C) and manganese (Mn), which are used to make steel. Structural steel has less than 0.30 percent C and 0 to 1.35 percent Mn. Note that adding carbon reduces toughness, while adding manganese improves toughness. In addition, other alloying elements used in making steel, and other effects such as degree of rolling, rate of cooling, and orientation of the Charpy V-notch specimen, can affect the toughness of a particular member. Thick plates and heavy shapes, which are worked less by rolling in the mill, are more likely to exhibit low toughness than similar thinner plates. For these reasons all critical members exposed to low temperatures should be tested for toughness prior to fabrication. In the absence of other criteria, the toughness requirements applicable to local highway bridge construction could be used. These requirements include the minimum amount of energy absorbed when tested at a specified temperature. In colder climates a lower testing temperature or a higher energy requirement is specified to ensure adequate toughness in service.

Even after specifying and getting an appropriate toughness value, detailing and workmanship can affect the performance of exposed steel. Reducing or eliminating cracks, stress concentrations, flaws, and notches reduces the likelihood of future cold-weather problems. If possible, avoid connections at exposed areas. If bolting, holes should be drilled or punched, not burned. If welding, welds should be inspected for flaws. If access holes are

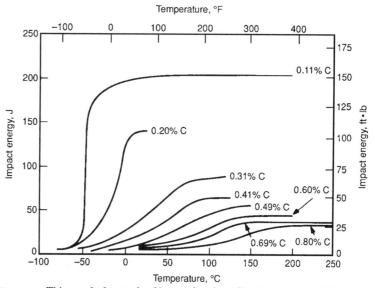

Figure 5.2 This graph, for steels of iron and carbon (C) only, shows that Charpy V-notch impact energy (toughness) varies with temperature and carbon content. Toughness increases and transition temperature drops (improves) with less carbon. (*From* Metals Handbook, *ASM International.*)

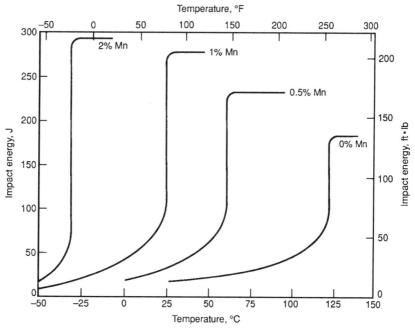

Figure 5.3 This graph of furnace-cooled steel alloys containing iron, 0.05% C, and varying amounts of manganese (Mn), shows Charpy V-notch impact energy (toughness) versus temperature. Note the well-defined transition temperatures. Energy rises and transition temperature drops with more Mn. (*From* Metals Handbook, *ASM International.*)

required, they should be cut in smooth curves, without corners, and ground notch-free. Copes and other openings should have generous corner radii, drilled or ground smooth.

Fatigue

Fatigue is the gradual reduction of material strength with cyclic load. An extreme example of this is breaking a wire or beverage can by bending it back and forth. While steel is affected by fatigue, the load cycles created by daily temperature swings are few enough in number (10,000 to 30,000 cycles over the building's useful life) that this is not a significant issue in exposed structures in buildings, as long as the steel is operating at a reasonable allowable stress. Where other cyclic loads occur, such as oscillating machinery, highway traffic, or sustained wind-driven vibrations, fatigue must be explicitly considered. For example, wind-driven oscillations of cables, stacks, and poles can reach millions of cycles in just a few months.

Fire Resistance

At the exceptionally high temperatures of a fire, there are also significant changes in the properties of steel. The strength of steel as well as its modulus of elasticity are affected. Figure 5.4 shows the decrease in the strength of a typical steel member as the temperature rises. Note the dramatic loss in strength starting at 315°C (600°F). The elastic modulus also drops above 315°C, potentially leading to premature elastic buckling of columns. A key question in the use of exposed steel is whether it requires fire protection. Most building codes do not require fireproofing where exiting is fast, as in low buildings with numerous windows and doors, or where distance from flames will keep temperatures low, such as at high ceilings in factories or gyms. Manufacturers are developing new types of steel with improved high-temperature performance. For example, Nippon Steel's SM50A-NFR maintains its yield point, tensile strength, and elastic modulus for about an additional 100°C (180°F) as compared to conventional steel. This can permit exposing the steel where the temperature is below 600°C (1100°F), or reducing fireproofing to a thin troweled coating which preserves the crisp lines of the members. Note that thermally induced stresses, deformations, and their effect on stability must be investigated if such elevated temperatures are anticipated.

If fire protection is required, methods can be used to insulate the steel or to slow its temperature rise in order to permit firefighters

Figure 5.4 High temperatures cause a drop in yield and tensile strength of structural steels as temperature exceeds 315°C (600°F) (*From* Steel Design Manual (1981), *Chap. 1, courtesy of R.L. Brockenbrough & Assoc. Inc.*)

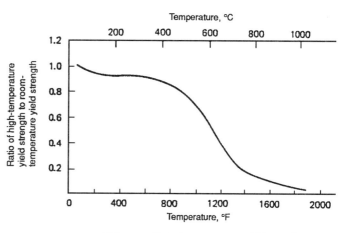

(a) Average effect of temperature on yield strength

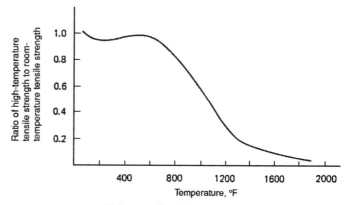

(b) Average effect of temperature on tensile strength

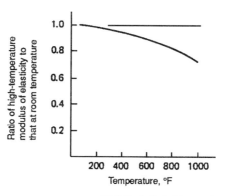

(c) Typical effect of temperature on modulus of elasticity

to control the blaze before collapse. In general, these methods would take the form of sprayed-on fireproofing, concrete encasement, or an enclosure of masonry or drywall. When it is desired to expose the form of the steel structure, there are other possible approaches. One modern coating looks like rough textured paint, but foams up when heated, protecting through insulation. Another coating sublimes (goes from a solid to a gas, like dry ice), absorbing heat in the process and cooling the steel. These intumescent and subliming coatings are generally used in weather-protected locations. Check with the manufacturers for possible outdoor application.

Heat sinks, using water or concrete inside tubular members, can actively carry away heat or increase thermal mass to slow the temperature rise. The U.S. Steel headquarters building in Pittsburgh, Pa., is one notable example, with a network of watertight steel box columns, expansion tanks, and pressure relief vents (see Chap. 3., Fig. 3.34)

Exposing steel in special cases requires special solutions. More and more often today fire protection is addressed analytically rather than by applying traditional approaches.

Fire modeling was used by the Chicago and London offices of Skidmore, Owings & Merrill (SOM) in developing the fire protection system for the exposed steel parabolic tied arches that carry a 10-story building 256 ft across active railroad tracks in London. The building, the centerpiece of the 14-building Broadgate development, has two interior arches and two that are exposed, all seven stories high (see Fig. 3.15).

Hal Iyengar, SOM partner and director of structural engineering, stated that with the interior of the building fully sprinklered, it was possible to examine exterior fire protection on a rational basis from fire loads in the building. A severe design fire was developed using a fire load of 30 kg/m^2 (6 lb/ft^2), which was determined to cause a fire equivalent to a 1 ½-hour fire resistance. A more severe fire loading of 60 kg/m^2 (12 lb/ft^2) was used to represent a storage room that can be placed anywhere along the exterior window wall—a 2-m- (6.5-ft)-wide gallery at the perimeter which permits the exterior structure to be exposed.

Heat transfer was then calculated into the steel structural elements to determine the maximum steel temperature for the given fire. After maximum steel temperatures were determined, a structural analysis was performed of the entire three-dimensional structural system at these temperatures to determine the forces and building deformations caused both by thermal expansion and by changing elastic properties. This ensured that steel temperatures were well below the critical level and that the entire structural system was sound and stable.

In general the temperature rise for steel was limited to 550°C (1022°F). Without any fire protection, steel temperatures of 675 to 704°C (1300°F) were found for the arch and hanger-column members, which were found to be unacceptable. One alternative considered was flame shields formed of sheet steel and attached to main members, but separated from them either by a small air gap or by a noncombustible insulation made of ceramic fibers. But these flame shields, while technically feasible, did not meet aesthetic requirements and involved concerns about durability, corrosion protection, and longterm maintenance. Instead, the design team chose a fire-resistant glass window wall, which was equivalent to having a fire-rated barrier between the fire load and the exposed steel.

To create the dramatic exposed steel columns and vaults on the interior of the United Airlines Terminal at O'Hare International Airport in Chicago, designers there also used an analytic approach to gain the approval of code officials, as described in a case history at the end of this chapter.

Corrosion Resistance

Corrosion, the gradual wearing away of material by chemical action, is a key factor to consider in exposing steel. Exposed structural members will certainly contact the air and thus be subject to oxidation. Aluminum forms a protective film of aluminum oxide when exposed, but when oxygen chemically combines with a ferrous (iron-based) material in oxidation, rusting or corrosion results. The oxides of iron formed in air or water are soluble in water and thus can be washed away. In addition, iron oxides take up many times the volume of the original iron, creating a swelling or wedging action, which causes further damage. This process is affected by temperature and especially by moisture, which particularly accelerates the process. Less moisture accumulation means less corrosion, an important consideration in detailing exterior building elements. Connections, changes in section, and fenestration details that cause accumulation of standing water or even just a film of moisture must always be avoided. If a pocket is unavoidable, it must be detailed to drain through the use of weep holes. Weep-hole size, quantity, and placement should consider the possibility of clogging, the accessibility and likelihood of cleaning, and the routing of weep water down inconspicuous paths such as reentrant corners and reveals (Fig. 5.5).

Even when it doesn't pond, water carries impurities and that

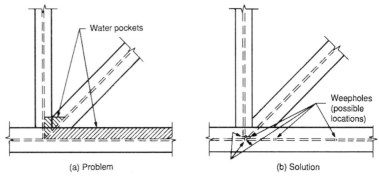

(a) Problem   (b) Solution

Figure 5.5 (*a*) Exposed structures must be detailed to avoid trapping water in pockets. (*b*) Where using closed sections or altering member orientations is not practical, generous weeps and snips can allow puddles to drain.

everpresent commodity called dirt, which can aggravate oxidation. Dirt can hold moisture for long periods of time, and water combined with dust and impurities in the air can also cause staining on a building. Exposed structures should be designed to minimize retention of water and dirt and to maximize runoff.

With the exception of weathering steel, all exposed steel structures must be protected from corrosion. Weathering steel, a product which became popular in the 1960s, is specially formulated to create a tightly adhering, weather-resistant oxide coating or patina. By avoiding the flaking nature of conventional rust, loss of material is greatly reduced (Fig. 5.6). However, even with weathering steel there are conditions where moisture can become entrapped and corrosion can be a problem. Experience in recent years has shown that running water, as under bridge scuppers or at curbs, can remove that protective patina through erosion and can cause harmful loss of steel. In addition, standing water, such as on flat surfaces and in unsealed plate-to-plate contact surfaces, causes uncontrolled oxidation with the same destructive cycle of swelling, flaking, and new oxidation as experienced by conventional steel.

In one building, exposed bolted weathering steel connections consisted of pairs of long butt plates to match the height of the deep members. Evenly spaced through-bolts held the plates face to face. Because relatively few bolts were needed to carry the design forces, the spacing of the bolts was several times greater than the usual 75- to 100-mm (3- to 4-in) gauge. This permitted slight gaps in plate contact to occur between bolts, and these gaps admitted water. Over time the buildup of rust forced the plates apart, failing the bolts in tension. In this building, accessible connections permit

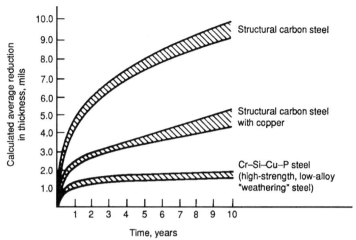

Figure 5.6 Comparative corrosion of steels in an industrial atmosphere can vary considerably. (*From Larrabee, 1953; © NACE.*)

maintaining structural safety by regular inspections and the replacement of failed bolts. A special penetrating sealer has also been applied to reduce water infiltration. In less accessible locations preventive measures could include using the tightest practical bolt spacing to minimize gaps, preventing rust on contact faces by prepainting them using zinc-rich coatings, sealing the edges of such joints (though this places a maintenance burden on the building owner), or shielding the joints from the weather.

In another building, weathering steel metal deck at parking levels trapped the salt-laden water, which worked its way down through normal hairline shrinkage cracks in the concrete slab fill and wearing surface. The deck rusted completely through where the water stood longest—at the deck troughs at the midspan low points, which occur naturally due to deck deflection under the wet concrete construction load. In this case the metal deck was being used compositely with the concrete, and was acting as slab tension reinforcement, so the loss of midspan deck steel required complete deck reconstruction. Metal deck should not be used as tension reinforcing for slabs exposed to chronic wetting. In such cases reinforcing bars should be provided, and metal deck, if used, should be considered as only temporary formwork with the understanding that it may corrode and need to be removed if it loosens in the future.

At existing highway bridges throughout the snow belt, state highway departments have addressed the issue of running water by painting weathering steel girders for several feet to each side of expansion joints and scuppers. This less than attractive approach can be avoided in new construction by minimizing joints, providing double gutter systems at joints, and diverting water away from the steel in a positive fashion.

The oxide coating which provides the protection for weathering steel takes time to form, from a few months in an urban industrial environment to a year or more in unpolluted rural areas. During that time rust is forming and washing off. In addition, some rust loss occurs all during the life of the structure.

Because of these staining characteristics of weathering steel, its use in exposed structures has become much less popular than when it was first introduced. Good practice calls for channeling potentially rust-colored runoff directly to drains, but this can be difficult to achieve. For example, runoff from overpass bridge girders can cause stains on pavement below, wash from spandrel beams can stain glazing below, and wind currents at taller buildings can scatter rusty droplets long distances during rainstorms.

The key to resisting corrosion, other than going to weathering steel, is a top-quality paint system. To provide durability and long useful life, paint thickness is important, since the paint itself will erode with time. And to avoid undermining and premature paint failure, excellent preparation is vital. As any house painter can tell you, the best paint in the world will not adhere to a dirty or weak substrate. Therefore before painting, exposed steel members should be cleaned to a higher standard than steel that is to be concealed, receive only primer, or receive fireproofing.

The Steel Structures Painting Council (SSPC) provides detailed specifications for the cleaning of structural steel which is to receive a finish surface, and recommendations as to their applicability. The degree of cleaning varies from surface preparation 1 (SP1; solvent cleaning using liquids or steam to remove grease and dirt) to SP5 (white metal blast cleaning). Numbers do not represent a scale of relative degree of cleaning. For architecturally exposed steel, which requires a top-quality finish, SSPC-SP6 (commercial blast cleaning) will remove dirt, grease, loose rust and scale, and at least two-thirds of all visible residues. By contrast, SSPC-SP3 (power-tool cleaning), removing loose dirt, rust, scale, and old paint, may suffice at exposed but not featured members, such as exposed floor beams, or at interior visible members. SSPC-SP2 (hand-tool cleaning) may suffice at members not exposed to view. The degree of cleaning could be linked to the designation of *architecturally exposed structural steel* members on the contract drawings, as described in the AISC code of standard practice.

Exposed members should then be covered by a primer, an intermediate "tie-in" coat (if required by the chemistry of the paint system used), and usually two finish coats, at least one of which is field-applied to give a uniform overall appearance. The term "paint system" is used deliberately here because coatings have advanced a long way from red lead and oil paints, and the chemistry of all the paint layers must be matched to each other, to the type of exposure anticipated, and to the application facilities available. These may be indoor or outdoor shops, may be heated or at ambient temperature, may include a curing room for baking or forced curing, and may involve electrostatic spray or brush application. Consideration of all these factors is crucial to final performance. Table 5.1 provides an overview of various coating types. Combining these coatings into systems requires care and should include input from coating manufacturers.

Wind. Wind aggravates other conditions as it drives water into structures. Even when surfaces are vertical, one must consider the effect of water hitting those vertical surfaces under pressure. In the design of wall systems, this phenomenon is considered through testing methods that simulate wind-driven rain. Such testing applies to exposed steel facade systems and to the interaction of conventional facades and exposed structures at joints and penetrations.

Facades are designed two different ways to resist water infiltration—as barriers and as pressure-relieved rain screens. Barrier wall systems have a single line of defense for both water and wind pressure. Rain screens use two layers—the inner layer resists pressure, while the outer one screens out most of the water. Since the dead air space between layers is at the same pressure as the exterior, any water entering it can easily be weeped out.

The American Society for Testing and Materials (ASTM) Standard E547, "Test for Water Penetration of Exterior Windows, Curtain Walls, and Doors by Cyclic Static Air Pressure Differential," specifies a test procedure in which a facade system is built as one wall of a sealed chamber, a partial vacuum is created in the chamber, a uniform water flow is provided by an array of spray heads outside the chamber, and observers in the chamber look for uncontrolled water inflow. Designers often go beyond this test and cycle through large positive and negative pressures (with no observers in the chamber, for safety) so that the structural integrity of the barrier wall is checked, and the water test can be repeated to check for damage to seals. Some testing laboratories are equipped to go even further and simulate hurricane conditions, using spray driven by airplane pro-

---

Table 5.1   Coating Comparisons

Coating	Strengths	Weaknesses
Cure by solvent evaporation		
Acrylic latex	Exterior durability Flexibility Fast drying Nonflammable	Abrasion resistance Chemical resistance Solvent resistance
Oil-modified alkyd	Flexibility Fair weathering High solids Penetration	Abrasion resistance Chemical resistance Solvent resistance Slow drying
Vinyl	Chemical resistance Fast drying Good weathering Flexibility Water immersion	Low solids Solvent resistance
Cure by coreaction		
Epoxy	Adhesion Chemical resistance High solids Solvent resistance Immersion service Abrasion resistance	Color and gloss retention Limited pot life
Coal-tar epoxy	Adhesion Chemical resistance High solids Solvent resistance Abrasion resistance Immersion resistance	Color availability Sunlight resistance fair Critical recoating time Limited pot life
Water-borne acrylic epoxy	Low odor and toxicity Adhesion Fast drying Recoatability Scrubbability Chemical resistance Solvent resistance	Initial moisture sensitivity Floor traffic resistance Medium-level solids
Moisture-cured aromatic urethane	Chemical resistance Adhesion Abrasion resistance Rapid cure Low-temperature cure Solvent resistance One-component formula	UV stability Initial moisture sensitivity Color availability
Aliphatic urethane	Excellent durability Chemical resistance Solvent resistance Abrasion resistance	Limited pot life Moisture sensitivity during curing

Table 5.1 Coating Comparisons (cont.)

	Galvanic	
Inorganic zinc-rich coatings	Galvanic protection Abrasion resistance Solvent resistance High heat service	Fair alkali resistance Poor acid resistance Touch-up and recoating difficult No brush or roller
Organic zinc-rich coatings	Galvanic protection Abrasion resistance Mild chemical exposures Rapid recoatability Application flexibility Good adhesion	Depending on binder: Moderate heat service Poor solvent resistance
Hot-dip galvanizing	Galvanic protection Abrasion resistance Solvent resistance High heat service Excellent adhesion	Must be applied at galvanizer facility Pickling and heat during dip can cause hydrogen embrittlement at welds and bends Heat during dip can cause distortion of members

SOURCE: Adapted from *TnemecTopics*.

pellers. Finally, careful observation during disassembly of the wall can reveal unsuspected leakage paths and water-holding pockets. Such a test can be valuable in pointing out weaknesses of a weatherproofing system and permitting the development of satisfactory solutions before construction.

Coefficient of Thermal Expansion

Like all materials, steel is subject to thermal movement. That is, even though no other properties may be affected, the material will expand and contract based on variations in temperature. The coefficient of thermal expansion of structural steel is 11.7×10^{-6} (11.7 millionths) mm/mm/°C (6.5×10^{-6} in/in/°F), which is close to the coefficient of expansion of concrete (9.9 millionths/°C, or 5.5 millionths/°F), and the reason they are able to work together in composite forms such as reinforced concrete and concrete-encased steel without cracking each other apart.

Temperature Variations. In an exposed design, movement must be considered. While exposed steel columns may be experiencing a surface temperature variation from –35 to 70°C (–30 to 160°F), they may be rigidly attached to the interior of the building, which is being maintained at a constant temperature of about 21°C (70°F). Differential movement results. For example, an exterior column on a 60-story building, exposed to just a 39°C (70°F) change in temperature, could experience a change in length of 100 mm (4 in).

The temperature variations that are simply a function of geography can produce considerable changes in the behavior of steel. In New York City, for example, the typical ambient temperature may vary from –18 to 38°C (0 to 100°F). For Arizona, it may vary from 0 to 46°C (30 to 115°F), while in certain areas of Alaska the temperature may drop to –51°C (–60°F) and rise only to 27°C (80°F).

But more important than the ambient (air) temperature is the actual temperature of the material itself. Steel will absorb heat from radiant energy. The temperature of an exposed steel member will be a function of ambient temperature, exposure to sunlight, color and reflectance (glossiness of surface), time of day, mass of the member, and ability for heat to flow to shaded or cooled surfaces. It is not uncommon to find ambient temperatures on a sunny day in the range of 21°C (70°F), but actual steel temperatures to be much higher, possibly 32 to 38°C (90 to 100°F). As a rule of thumb, consider that the steel temperature of a white-painted member may be about 14°C (25°F) greater than the ambient temperature. With darker colors and protection from cooling breezes it could be much greater. In very windy areas, the air constantly transfers heat from the steel and will tend to keep the steel cooler than in areas where there is less wind.

The American Society of Heating, Refrigerating and Air-Conditioning Engineers (ASHRAE) *Handbook of Fundamentals* (1985) presents design ambient maximum and minimum temperatures for hundreds of locations in the United States and Canada. It also provides design surface temperatures for the walls and roofs of buildings to be cooled, and adjustments for orientation (such as facing North), latitude, time of year, time of day, and surface color. However, the surface temperatures provided or implied, based on the Sol-Air method or the cooling-load temperature-difference method, are not necessarily actual surface temperatures since they were developed only as convenient ways to quantify the heat load from the sun (Sol) and the outside air for air-conditioning design. A site-specific study, using established principles of physics, could better determine likely variations in steel temperature.

Structural Behavior. When only part of the structural system is

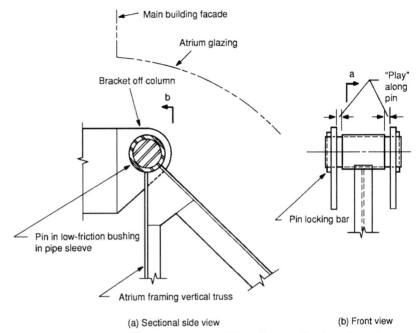

Main building facade

Atrium glazing

Bracket off column

b

a

"Play" along pin

Pin in low-friction bushing in pipe sleeve

Pin locking bar

Atrium framing vertical truss

(a) Sectional side view

(b) Front view

Figure 5.7 Joint on the Northwest Atrium Building, Chicago, Ill., allows for atrium thermal movement. (*a*) Sectional side view. The pin is on a column bracket, the sleeve is on the atrium glazing support truss. (*b*) Front view. "Play" along the pin permits thermal movement of the atrium relative to the building.

exposed to temperature changes, certain thermal effects have to be considered for the overall structural behavior of the building. When the outside temperature varies but the inside temperature is essentially constant, relative elongation or contraction between exterior and interior members of the structure can have a significant effect on stresses as well as deformations. A building can experience both local and overall effects due to the differential thermal condition.

One way to deal with these effects is by isolating the exposed elements from the rest of the structure. For example, Northwest Atrium is a 40-story office tower with seven-story glass-covered atria on two sides. The atrium support framing is isolated from building movements, and vice versa, by providing rotational pins which allow rotation and along-pin sliding movements (Fig. 5.7).

Another approach is to anticipate and design for the induced movements and forces while leaving the exposed elements connected to the building. However, this can have major implications. On highrise buildings, for example, the effect of exposing portions of the structure can make a pronounced difference in the structural behavior of the building, drastically affecting frame design and cost.

The most common lateral (wind) bracing system for steel mid- to high-rise buildings is the braced core. Where a core becomes too tall or slender to provide efficient bracing, outrigger trusses, brackets, hats, or girders are often used to connect the core to perimeter columns, and thus stiffen the core and resist overturning. These members can cause problems where thermal motion occurs. For example, differential thermal motion (one side of the building warmer than the other) can cause the entire building to tilt (Fig. 5.8a). Seasonal variations leading to uniform thermal changes around the perimeter can cause all the perimeter columns to pull down or push up on the core by means of the outrigger trusses (Fig. 5.8b). With heavy column sections and stiff trusses, the forces involved can become large. If these forces are included in the design of the perimeter columns, core columns, and outriggers, the frame will be quite inefficient (heavier than otherwise necessary). Such a situation should cause the designer to rethink the bracing approach used and the impact of exposed members on the design.

Exposing one side of a braced core to the outside environment can lead to tilting caused by unequal thermal motion. For any given nonuniform temperature distribution, unequal strains (length changes) result. Difference in strain divided by distance over which the difference occurs defines curvature. Curvature times building height provides building top lateral movement. Thus the motion induced by unequal temperature on a 50-ft-wide steel braced core will be about four times larger than that for a 200-ft-wide tube-type exposed steel building (compare Fig. 5.8a and c). Lateral movement can be minimized by forces acting between opposed pairs of walls or cores (Fig. 5.8d).

A tube-type building has closely spaced columns moment-connected to deep spandrel beams on all faces, forming a "wraparound frame." Tube buildings will also exhibit overall tilting under nonuniform thermal conditions since relatively stiff exterior frames enforce compatible motions. The magnitude of the motion can be estimated by considering the tower as cantilever column with a variation in column strain, yielding a corresponding curvature. Rate of curvature times building height gives lateral motion (Fig. 5.8a).

For both the braced core and the tube-type systems, if floor beams between interior and exterior columns are reasonably long and pin-ended (normal double-angle, seated, or shear tab connections), seasonal variations in column length will pose little problem (Fig. 5.8e). Slight tilting of the floor will be the only effect, which

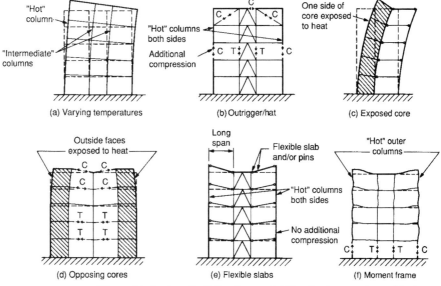

Figure 5.8 Thermal behavior of building frames. (*a*) Varying temperatures. If column temperatures and column strains vary, there is overall lateral building movement, but little differential movement across any one bay. (*b*) Outrigger/hat system. Perimeter columns connect to the core. Differential temperatures generate potentially large internal forces. (*c*) Narrow core with one side exposed. Temperature changes will cause overall lateral building movement greater than that for the full frame in *a*. (*d*) Exposed cores working in opposition. They will reduce lateral building movement but generate significant internal forces. (*e*) Flexible slabs/pinned beams. Exposed columns will exhibit differential movement, but will not induce forces. (*f*) Moment frame. Differential column movement will induce internal forces. (*Adapted from White and Salmon*, Building Structural Design Handbook, *John Wiley & Sons, Inc. 1987.*)

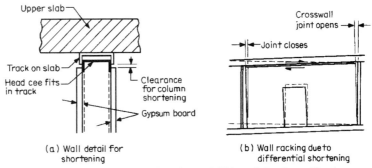

Figure 5.9 Interior partition considerations. (*a*) This wall detail provides for shortening of columns and racking of floors. Gypsum board can lap over the top track only if care is taken not to fasten board to top track. (*b*) Wall racking occurs due to differential column shortening. (*From White and Salmon, John Wiley & Sons, Inc. 1987.*)

must be addressed by the construction details used at heads and ends of interior partitions (Fig. 5.9). Conventional slabs on metal deck will usually have sufficient flexibility also. Where slabs are unusually thick, such slabs should be checked for induced forces and detailed to provide the required flexibility. When conventional frames with interior floor beam-to-column moment connections include both interior and exterior columns, thermal effects must be accounted for by considering imposed length changes (Fig. 5.8*f*). For moment-connected beams of normal length, spanning typical 30- to 40-ft bays, the moments and shears generated will be significant but not overwhelming (as in the case of outrigger trusses) because frame flexibility accommodates the differential movement with smaller induced forces.

CASE HISTORIES

Case 1

One Mellon Bank Center (OMBC), a 54-story, 726-ft-high office tower developed by U.S. Steel Realty in Pittsburgh, Pa., illustrates many of the points discussed in this chapter regarding exposed structural steel. At OMBC the building's steel skin is used as an exposed structural system to help stiffen the building and as its weatherproof facade. The design accommodates thermal movements, meets applicable fire protection codes, considers stress concentrations and toughness, and avoids corrosion.

Instead of anticipating and allowing for overall building movement, the approach to exposed structure used at OMBC was to isolate the exposed elements to permit unencumbered movement. Thermal movements are accommodated by means of flexible horizontal joints every two to three stories. Flexible plates permit skin panels to expand and contract, while still holding edges securely against shear-type movements so that the panels act as shear diaphragms or infills (Fig. 5.10).

The fire code was satisfied by treating the skin as a nonparticipating element for overall building strength and stability, and considering it as acting solely as a comfort-and-drift control feature. Thus the building was designed twice—with and without the skin participating.

The facade includes rectangular windows held by "zipper strip" rubber gaskets in holes torch-cut in the facade plates. Because sharp corners cause stress concentrations which can lead to local cracking or yielding, generously rounded corners were used. A local "teardrop" bulge of the opening edge permits rounded corners

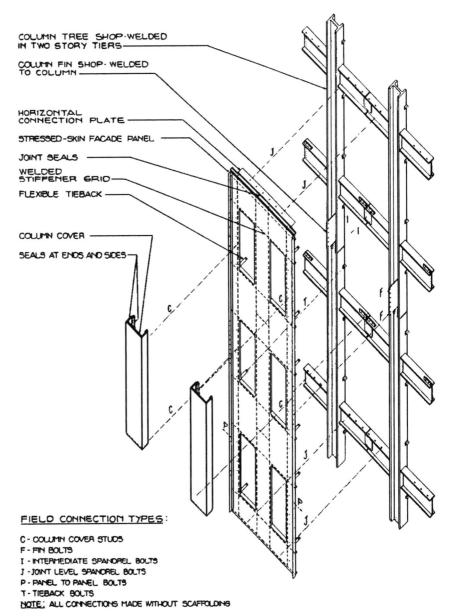

COLUMN TREE SHOP·WELDED
IN TWO STORY TIERS

COLUMN FIN SHOP· WELDED
TO COLUMN

HORIZONTAL
CONNECTION PLATE

STRESSED-SKIN FACADE PANEL

JOINT SEALS

WELDED
STIFFENER GRID

FLEXIBLE TIEBACK

COLUMN COVER

SEALS AT ENDS AND SIDES

FIELD CONNECTION TYPES:

C - COLUMN COVER STUDS
F - FIN BOLTS
I - INTERMEDIATE SPANDREL BOLTS
J - JOINT LEVEL SPANDREL BOLTS
P - PANEL TO PANEL BOLTS
T - TIEBACK BOLTS
NOTE: ALL CONNECTIONS MADE WITHOUT SCAFFOLDING

Figure 5.10 Connections between stressed skin and building frame at One Mellon Bank Center support the skin panels laterally and allow for thermal movement while transmitting shear forces. (*From White and Salmon, John Wiley & Sons Inc. 1987.*)

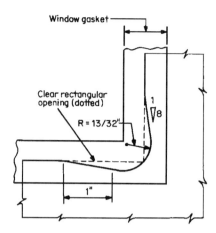

Figure 5.11 This "teardrop" corner detail is used at window openings in steel stressed skin panels to reduce local stresses while allowing space for the glazing itself.

without affecting the glass corners. The rubber glazing gasket covers this bulge (Fig. 5.11). The numerically controlled plasma torches used to cut openings provided a smooth, notch-free edge, so grinding was minimized. The thin plates used (6 to 8 mm, or 1/4 to 5/16 in) eliminated concerns for toughness because this quality is enhanced in the process of rolling the plate down from billets.

The facade details included seal welds at all seams and a positive pitch to drain at joints that could trap water, provided by weld profiles and body filler. The facade panels are protected by a uniform 2- to 3-mil cover of inorganic zinc-rich primer throughout and, at the exterior face, a 0.5-mil basic zinc chromate vinyl butyrate wash primer as a binder and two 2- to 3-mil acrylic urethane finish coats, all shop-applied. Note that controlling paint thickness and ambient relative humidity during application was important to the proper performance of this system. The finished building is shown in Fig. 5.12.

Figure 5.12 Exposed steel stressed skin panels add lateral stiffness to this building for occupant comfort.
One Mellon Bank Center, Pittsburgh, Pa.; Architect: *Welton Becket Associates*; Engineer: *Lev Zetlin Associates Inc/Thornton-Tomasetti.* (Photo: *Len Joseph.*)

Figure 5.13 This concourse and a sister one total 1 km (3300 ft) of building length. Arched steel framing meets three-pipe cluster columns at outside walls and four-pipe columns along the ticketing area.
United Airlines Terminal, O'Hare Airport, Chicago, Ill.; Architect: *Murphy/Jahn with A. Epstein*; Engineer: *Thornton-Tomasetti Engineers*. (Photo: *Charles Thornton*.)

Case 2

The United Airlines Terminal at O'Hare Airport, Chicago, Ill., exposes structure for quite a different purpose. Only a small portion of this steel is exposed to weather, but it is all exposed to view to recall the excitement of travel evoked by the great train-sheds of European rail stations. Two concourses totaling 1 km (3300 ft) in length provide vistas of white steel arched frames (Fig. 5.13). An adjacent ticketing area uses folded-plate trusses to span 37 m (120 ft) (see Fig. 8.1).

The steel in the United terminal would be exposed to constant pedestrian traffic, so safety, comfort, and ease of maintenance

Figure 5.14 Perforated steel is used in gate-side lounges also. Note one-pipe columns at the outside wall and four-pipe columns in the lounge. Fire studies considering building size, volume, and effect of sprinklers convinced code officials that the steel could be left exposed.
United Airlines Terminal, O'Hare Airport, Chicago, Ill.; Architect: *Murphy/Jahn with A. Epstein*; Engineer: *Thornton-Tomasetti Engineers.* (Photo: *Charles Thornton.*)

were addressed by using pipes and pipe clusters for columns. Rounded shapes avoided the sharp corners and pockets of wide flange members—an asset for safety, ease of cleaning, and longer paint life. Clustering small pipes to make larger columns enhanced visibility and the airy feeling desired. Clustering also led to a hierarchy of column types and consistent treatment of frame-to-column connection details. Lightly loaded beams at side lounges have one-pipe columns (Fig. 5.14), folded-plate truss columns have two-pipe columns (Fig. 5.15), concourse frames loaded at outside walls use three-pipe columns and, at double-sided loads, four-pipe columns (see Fig. 5.13). Where arch frames

Figure 5.15 Folded-plate trusses over the ticketing area are supported by these two-pipe columns exposed to weather. Toughness (fracture control) criteria for member steel and weld metal were similar to those required for highway bridges.
United Airlines Terminal, O'Hare Airport, Chicago, Ill.; Architect: *Murphy/Jahn with A. Epstein*; Engineer: *Thornton-Tomasetti Engineers*. (Photo: *Charles Thornton*.)

Figure 5.16 Half-domes terminate the concourse ends with an unusual twist—glazing tubes curve around as horizontal "latitude" elements while smaller sag rods run vertically as "longitude" elements.
United Airlines Terminal, O'Hare Airport, Chicago, Ill.; Architect: *Murphy/Jahn with A. Epstein*; Engineer: *Thornton-Tomasetti Engineers*. (Photo: *Charles Thornton*.)

are doubled up to change concourse height, five-pipe columns occur. The pipe motif was carried into the roof purlins, where round tubes with mounting tees provide biaxial bending strength, create minimal shadowing, and offer no ledges to trap water or dirt. Their strength also permitted unusual framing for the half-domes at concourse ends. Purlins wrap around the domes horizontally, like latitude lines, while their weight is carried by light sag rods arranged like longitude lines (Fig. 5.16).

Toughness was a concern for this project, both because some elements are exposed to Chicago's notoriously cold winters (see Fig.

5.15) and because even "indoor" elements were to be erected in winter. Charpy V-notch values consistent with local highway bridge design requirements were specified for main steel members and weld metal.

Although mostly enclosed behind a weathertight envelope of fritted glass, thermal movement was still an issue in these 1600- to 1700-ft long buildings. The tubular glazing support purlins were provided with telescoping joints and the wide-flange framing with link hangers every 90 to 120 m (300 to 400 ft). Analysis indicated that individual frames could accommodate local thermal effects.

The exposed steel columns and vaults would have lost all their impact if they had to be clad by fireproofing. Even the "high-tech" coatings discussed above would not have been satisfactory, as their thickness would detract from the desired crisp lines and they are not intended for surfaces experiencing frequent contact and abuse. Architect Helmut Jahn, structural engineers Thornton-Tomasetti, and fire protection consultant Schirmer Engineering Corporation had to convince local authorities that the design was safe, although it did not conform to local codes. But Chicago, like many other cities, now allows designers to demonstrate alternate means of compliance. At the United Airlines terminal, the design team produced specific test data demonstrating that the sprinkler system installed was capable of protecting the exposed steel. Also, an analytic fire model was developed, which showed that the size and volume of the building would prevent the development of critical temperatures.

CONCLUSION

The high strength and stiffness of steel makes it an ideal material for light, crisp, and elegant exposed structures. Steel is suitable for exposure in a wide variety of environments as long as toughness, fire protection, corrosion protection, and thermal movements are properly considered.

6 *Other Materials*

While concrete and steel represent the most commonly used structural materials, others can also be used in exposed structures to great effect.

WOOD

Wood is a material full of contradictions. The general public perceives it as "user friendly"—who hasn't done a little hammering, sawing, or whittling?—but engineers must recognize that designing for its use is more complicated than for other materials due to the directionality of wood fibers.

Most people think that wood connections are simple—a little glue and a few screws should do the trick—but designers must know the strengths of a wide variety of connectors when acting along, across, or at an angle to the grain. In addition, the designer's preference and ingenuity can determine whether the connections are straightforward and visible affairs of steel and boltheads (see Fig. 2.8), intricate and hidden (and more costly) assemblies of kerfs, recesses, and dowels, or somewhere in between.

Most people think of wood as an untreated, natural material—chop a tree down, saw it up, and use it—but some of the most useful forest products, plywood sheets, and glued laminated (glulam) beams are manufactured items containing as much engineering and fabrication effort as any concrete or steel member.

Most people think of wood as highly flammable, and this is true to a point. But the members used in heavy timber and glulam construction have a low surface-to-volume ratio and great thickness. These members are harder to ignite and only burn slowly, providing fire resistance without fireproofing. Thus, ironically, wood frames may be exposed in locations where noncombustible steel would require fireproofing. Fire-retardant treatments further extend wood's usefulness as an exposed material.

Most people think of wood as perishable and not suitable for monumental construction. But if protected from weather and kept at a relative humidity less than 30 percent, and below its fiber saturation point, wood will last indefinitely. Wood-framed temples in Japan are centuries old and still in good condition.

Most people think of wood finish as a low-tech item—just slap on some paint, varnish, or shellac. But the organic nature of wood affects the appropriateness of various finish treatments. Designers must establish the required exposure and desired performance before selecting a finish. Interestingly, the traditional finishes which show off wood's beauty are the least durable in sunlight. Clear finishes let ultraviolet radiation reach the wood surface. This causes wood fibers to deteriorate, undermines the support of the finish, and leads to crazing, cracking, and flaking of the finish.

Clearly, wood design requires specialized knowledge from sources such as those listed in the bibliography. In this chapter we will highlight items that may be of special interest in the design of exposed structures.

Structural Properties

The structural properties of wood relate directly to the direction, density, and continuity of the wood fibers. A cross section of wood from temperate climates exhibits rings due to alternating layers of low-density spring wood and high-density summer wood. Tropical woods, in contrast, are all summer wood. They often show no rings and are extremely dense and hard. Some species are so hard that nail holes must be predrilled and fabrication may require tools and processes more familiar to a metalworking shop than a carpentry shop.

Focusing on the temperate coniferous woods which are used in most North American timber construction (Douglas fir and southern pine), along-grain strength and stiffness (elastic modulus) is great since the bands of summer wood can act to carry load without involving the spring wood. The tubular nature of wood fibers also plays a role; just as a drinking straw can be stiff along its length but easily flattened, wood fibers crush more easily across the grain. In addition, cross-grain strength involves alternating layers of spring and summer wood, and spring wood sets the strength limit as the weakest "link in this chain." Cross-grain compression strength can be one-half to one-third of along-grain values, and cross-grain tension is frowned upon as too risky for sustained loads. As a result, one maxim of wood design is, "Load in compression or bending *parallel* to the grain." This affects the design and orientation of members and also their connections, as discussed later.

A natural material, wood contains continuous longitudinal hollow tubes which transport and store water in the living tree. The solid material around and between the tubes contains cellulose and lignin, which are nature's versions of synthetic plastics. Tubes, cellulose, and lignin can all gain and lose water. In doing so, the wood changes volume, particularly in the cross-grain direction. This leads to a second maxim of wood design, "Allow for cross-grain shrinkage." To do so requires careful thought and planning of details and joints.

Another result of being a naturally occurring material is that wood can serve as a food source. Fungus spores are always present in the air and are continually being deposited on horizontal surfaces. This form of life digests wood fibers. In the forest we admire those fungi known as mushrooms for their beauty, flavor, and role in nature's recycling system. In a building, related fungi which cause damage from rot are instead viewed as a menace to structural safety. Defense against rot can take two paths. Although people use the term "dry rot," this only means that the rotten wood was dry when observed, but must have been wet at some time. In reality fungal growth cannot occur if the wood is kept dry, below 30 percent relative humidity (RH) and the fiber saturation point. Thus the maxim, "Keep it dry." Good practice is to keep wood at 20 percent RH or less.

Crevices and gaps which occur at joints tend to trap water. Tubes exposed at member ends as "end grain" also absorb water faster and hold water longer than "side-grain" surfaces. For these reasons exposing joints to weather is never a good idea. Those picturesque covers over the wooden bridges of New England were provided solely to keep the joints dry! A second approach is to render the wood inedible by adding poison such as creosote or metallic salts. However, this approach must be used with care, balancing the effects of color, odor, touch, longevity, environmental impact, and cost.

Insects also view wood as a food source. In climates with cold winters the primary concern is subterranean termites. They eat the spring wood but maintain their nest in the earth; so, creating a barrier between soil and wood is an adequate defense. However, carpenter ants and a variety of tropical termites and beetles can both eat and live in wood. To deter these pests the use of treated wood is recommended.

Design Concerns

Design concerns for wood members include load duration, humidity, size, appearance, stress direction, and deflection control. Wood can resist loads of sudden or short duration, such as wind, earthquake, blast, or moving live loads, better than permanent or sustained loads. This can prove helpful in the event of unanticipated loadings. It can also improve structural efficiency in situations where the design live load will only rarely be seen, such as wall framing on a hurricane coastline or roof framing in a snow-free area. Conversely, if loads may be of long duration, such as snow load in a ski resort, allowable stresses must be reduced when considering these loads.

Humidity also affects allowable stresses—generally higher humidity means lower stress limits. This should be addressed where framing is exposed to ambient atmospheric humidity, such as at an open-air pavilion or bandshell. It would also be a consideration in high-humidity spaces such as swimming pool enclosures and industrial facilities.

The member size also affects allowable stresses, as larger members are more likely to include detrimental splits. This is reflected in allowable stress tables and footnotes, but should be especially noted in exposed structures, since heavy timbers are often used in such situations.

Directly related to humidity and member size is appearance. Wood members are cut from trees soon after the trees are cut down and while the wood is still green with a high moisture content. The wood is then dried by stacking and storing or by heating in a kiln. At this time the wood shrinks relative to the growth rings, both radially (toward the center of the tree) and circumferentially (along the ring lines). Shrinkage creates a cross-grain tension which leads to checks and splits. The drier the wood is in use, the more shrinkage will occur. The more massive the member, the more restraint against shrinkage and consequently the more splitting will occur. Allowable stresses are adjusted to anticipate strength loss from splitting. Therefore it must be expected that heavy timber construction will have a rustic appearance.

Where a more finished appearance is desired or where high strengths are required, glulam members can be used. Glulam members have thin pieces (laminations) which are individually kiln-dried before assembly (gluing), so shrinkage restraint and shrinkage cracking are minimized. Individual laminations can also be selected to provide an attractive knot-free appearance at all exposed faces.

The axial nature of wood fibers requires that the stress direction be considered carefully for a safe, efficient structure. For example, whenever possible, bearing details should have forces acting parallel to the grain rather than perpendicular. Where grain orientation is dissimilar between two pieces, a bearing plate may be required to spread the load. A number of clever details have been developed over the years with this in mind. Structural systems which resist loads primarily through axial compression are especially suitable for dramatic wood framing. Trusses can mix wood compression chords and steel rod or cable tension chords and diagonals, as at Wolf Trap Farm Park (see Fig. 2.5). Early trusses had chords and compression diagonals in wood, with tension verticals in wrought iron or steel. Wood-to-wood joints had simple bearing details, which were held snug by pretensioning the verticals through turnbuckles or threads and nuts. Wood can be used in tension, but gradual loss of strength can cause delayed failure, so it should be used at low stresses, supplemented with steel rods, or sandwiched with hidden steel flitch plates.

Arches act primarily in compression, but the difficulty of assembling a series of small pieces to follow a curve made them impractical until recently. With the development of glulam construction, thin wood-strip laminations can be held in jigs to follow any desired curve. Joints at lamination ends can be butted, scarfed (tapered), or finger-jointed to transfer stresses, and joint locations can be staggered among the laminations to avoid weak points. Once the laminations are glued together, the curved shape will be permanently retained (Fig. 6.1).

The ability to provide selected laminations on outer faces gives glulam members excellent flexural strength. This has made the development of standardized rigid frames practical and economical. The popular Tudor frame offers three attractions (Fig. 6.2). First, the outer profile has straight lines, which provide support for a simple economical gable roof and straight vertical sidewalls. Second, the curved inner profile provides an attractive, softer appearance. For tall frames with short spans the effect recalls Gothic arches and is popular for churches. Third, the difference between inner and outer profiles provides stiff, strong "knees" at the eave line to resist the moments from three-hinge frame action. Simple, economic bearing details provide the three hinges (at the two bases and at the peak). Unanticipated foundation movements are less significant for three-hinge frames than for two-hinge or hingeless frames.

Members acting primarily in flexure must also be designed with directionality in mind. Of course, such members should have grain parallel to the direction of flexural tension stress whenever possible. This concern can be particularly well addressed by using glulam members where lengths, heights, tapers, or desired allowable stresses are outside the range of available sawn members.

A frequently forgotten trouble spot is at dapped (notched) ends of beams. If the bottom of a beam is notched, shear in the body of the beam will be collected at the shallower dapped end through

Figure 6.1 This building interprets the classic European train shed in a unique way with a mix of exposed structural materials: rusticated cast-in-place concrete, glued-laminated wood arches, wood posts, wood rafters, and steel ties and hangers. **Back Bay Station, Boston, Mass.**; Architect: *Kallmann, McKinnell and Wood/Bond Ryder*; Engineer: *Weidlinger Associates*. (Photo: *Steve Rosenthal*.)

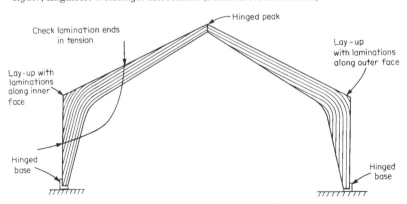

Figure 6.2 This Tudor frame in glulam wood shows two different lay-up options.

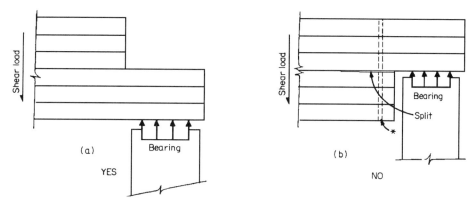

Fig. 6.3 Dapped (notched) ends of wood beams require particular care. (*a*) If the reduced depth of beam is adequate for shear, a notch at beam top can be tolerated. (*b*) A dap at beam bottom is an invitation to serious splitting. The through-bolt denoted by * may help to reduce splitting but it is better to avoid this situation entirely.

cross-grain tension near the notch. This can cause splitting (Fig. 6.3).

Directionality of stress is particularly important when designing connections. Loads on a connector might act in bearing parallel to grain, bearing perpendicular to grain, withdrawal perpendicular to grain (side grain), withdrawal parallel to grain (pulling from end grain), or any combination of these. The Hankinson formula provides a way to interpolate strengths for loads acting at an angle between known capacities parallel and perpendicular to the grain. Side-grain withdrawal values are tabulated for nails and screws, while end-grain withdrawal is not recommended at all.

The old familiar standbys, nails and screws, function by shaft bearing and friction. Their narrow shaft widths limit their capacities. Steel through-bolts can have larger diameters to increase bearing areas, but too many bolt holes can reduce a member's strength. Shear plates and split-ring connectors fit steel rims or rings into precut grooves to provide large bearing areas within shallow depths to minimize the adverse effect on member area. Small bolts then suffice to transfer forces between connectors or to hold wood pieces together (Fig. 6.4). Toothed grids and gang-nail plates are pressed into place as a way to mechanize the lighter connections of shop-fabricated trusses (Figs. 6.5 and 6.6). Where multiple connection devices are needed at one location, special care is required to avoid cross-grain restraint. At a connection, a wood

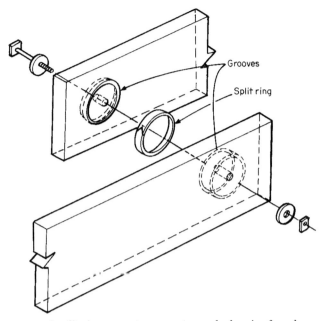

Figure 6.4 A split-ring connector uses a tapered-edge ring forced into matching grooves precut in the wood using templates. Load is transmitted by wood-to-ring bearing. The bolt acts to hold the pieces together. Common in heavy truss construction, it can transmit heavy loads with a minimum of exposed hardware. *Note:* Connections should be laid up symmetrically. The unsymmetrical layout shown here is for drawing clarity only.

Figure 6.5 A toothed grid connector functions like multiple nails, driven by forcing wood members together in a press. *Note:* Connections should be laid up symmetrically. The unsymmetrical layout shown here is for drawing clarity only.

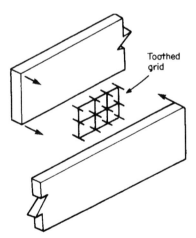

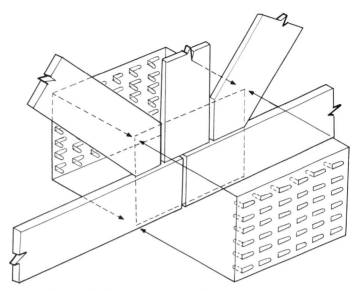

Figure 6.6 A gang nail plate serves as a combination gusset and nails, driven by forcing plates into members in a press. This is the most common connection for light residential floor and roof trusses.

member should be expected to contract toward a bearing point, so bolts or nails, if needed, should be located near the bearing point or should be detailed for free movement in the event of shrinkage. Figure 6.7 shows a moment connection where the top plate is free to move downward if shrinkage occurs. Figure 6.8 shows a few variations of beam end support details. A similar range of exposed, semiconcealed, and concealed details could be developed for virtually any connection condition, limited only by the designer's ingenuity. The American Institute of Timber Construction Standard AITC 104-84 provides guidelines and examples of a wide variety of such connection designs.

Deflection control with wood requires the recognition of three facts. First, since wood is so much lighter than steel, the temptation is to assume that it should also be built shallower than steel. However, the ratio of allowable stress to elastic modulus is similar to that of steel, which means that similar span-to-depth ratios should apply if desired deflection criteria are the same. Second, wood moves with changes in relative humidity. Therefore provision should be made in finishes and enclosures to accommodate these movements. Third, wood creeps and sets under long-term sustained loads, as can be seen in the sagged floors of colonial-era houses and the now

Figure 6.7 In this photo of moment-connected glulam beams, vertical shear is taken in bearing on brackets, while vertical bolts and a separate top moment plate (not shown) permit beam vertical shrinkage. The top plate and bolt heads are accommodated by wood fill strips. (Photo: *Len Joseph.*)

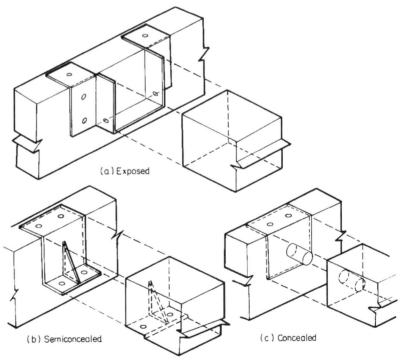

(a) Exposed

(b) Semiconcealed

(c) Concealed

Figure 6.8 Wood connections for simple shear can take a variety of approaches: (*a*) Exposed steel saddle. (*b*) Semiconcealed steel bracket. (*c*) Concealed steel pintle on recessed steel bracket (check beam for splitting as at a dap, and provide for rollover stability).

empty mill buildings of the early 1800s. So if sustained load is a major portion of total design load, and if long-term deflections are of concern for reasons of appearance, usability, or drainage, providing extra stiffness in the design is recommended.

ALUMINUM

With its natural corrosion resistance, aluminum would seem to be an ideal material for exposed structural applications. Indeed, aluminum often serves a vital local structural function in the wind-resisting mullions of curtain walls and the safety railings of stairs and balconies. Its corrosion resistance stems from the fact that aluminum oxides adhere tightly to the parent metal, protecting it from further oxidation in most applications once a thin oxide layer forms. However, the oxide layer is affected by alloying elements in the aluminum, so for best appearance some finish is usually required. An anodized finish preserves what we think of as an aluminum "look." Where greater protection is desired, paint can be used.

However, aluminum has not caught on as a primary structural material except for specialized applications. It hasn't been for lack of trying. There has even been a demonstration all-aluminum highway overpass bridge. The main obstacle is economics—aluminum has roughly one-third the density, one-third the strength, and one-third the elastic modulus of steel, so members would be of similar depth and strength per unit weight. Special high-performance aluminum alloys can do somewhat better than conventional steel in strength per unit weight. Even so, the price per unit weight is higher for aluminum than for steel. Aluminum has the advantage where maintenance costs are significant and where intricate extruded shapes or sharp edges are required for appearance or for assembly. It also is used, of course, where a certain thickness or volume is required by stability, and the lower density gives lower structural weight, as in aircraft frames.

COMPOSITES

The term "composites" is used here to include members which are composed of one or more high-performance materials embedded in a low-performance bulk matrix. The most widespread such composite is glass-fiber reinforced plastic (GFRP), known as fiberglass. Where structural demands are multidirectional and of low magnitude, such as for recreational boat hulls, mats of randomly oriented chopped glass fibers are impregnated with plastic resin. Where greater strength is needed, the mats can be woven, with long fibers running in two directions. Long, thin members can be made as bars, rods, and Is by the pultrusion process. Glass fibers are pulled off spools, through a resin bath, and then through a die of the desired shape, which forms the member and removes excess resin before hardening. The unaxial orientation of fibers gives the member excellent axial and flexural strength and stiffness. Where multidirectional strength is required, glass fibers can be looped around a jig in the desired pattern, then placed in a resin-filled mold for casting. This process is used to produce grids and gratings.

The main advantages of fiberglass are corrosion resistance, light weight, and nonmagnetic, nonmetallic properties. The resistance to chemical attack makes it popular for access grating and light-load applications in chemical plants. The light weight and "transparency" to electromagnetic radiation are big advantages when enclosing and protecting radio and microwave transmitters and receivers. The light weight and high relative strength and stiffness compared to other fibers make fiberglass fabric a practical choice for tents and air-supported roofs (see Figs. 3.23 and 8.18).

Fiberglass drawbacks include cost, strength, brittleness, and ultraviolet damage. Current manufacturing processes are low-volume, multistep, semicustom operations. Given the need for matching fiber layout with anticipated forces, this is unlikely to change very soon. Without mass-production techniques costs will remain higher per unit strength than for other structural materials. Glass fibers themselves are the limiting factor in GFRP strength. High-performance fibers such as DuPont's Kevlar would increase strength, but would not increase stiffness in proportion. Boron and carbon filaments have been used as stiff, strong reinforcing bands in very limited critical structural aerospace applications, but are not yet cost-effective for building construction.

GFRP can have good strength, but making connections can be particularly tricky due to brittleness. Structural steel has good ductility, accommodating local stress concentrations by yielding and deforming plastically. Since GFRP does not behave this way, all connections must be analyzed to determine and eliminate unacceptably high stress points by reshaping or reinforcing the members. Where exposed to ultraviolet rays, the bulk binder of GFRP can be degraded, damaging strength and finish over time. One way to control this is by adding opaque dye to the resin, limiting ultraviolet ray penetration. However, this may limit the selection of available colors and finishes.

GLASS

To maximize visual lightness, some designers have used structural glazing where solid glass bar mullions laterally brace glass windows. For light loads, connections can be made via silicone sealant. For heavier loads, mechanical connections have been used in the form of bolts and brackets cushioned from glass contact by rubber sleeves, bushings, and washers. This system is using glass in bending, primarily to resist wind. Clearly this approach, which emphasizes aesthetics over economics, can be appropriate at highly visible locations (Figs. 2.37, 6.9, and 6.10).

Figure 6.9 This view looking outward shows steel hardware used in structural glazing. Small squares restrain the exterior glass and larger bands splice the vertical mullions.
599 Lexington Avenue, New York, N.Y.; Architect: *Edward Larrabee Barnes/John M. Y. Lee & Partners*; Engineer: *Thornton-Tomasetti Engineers.* (Photo: *Len Joseph.*)

Figure 6.10 An "invisible" exposed structure. Glass is the structural element here, as glass vertical mullions resist lateral load on this atrium wall.
U.S. Fidelity and Guarantee Headquarters, Baltimore, Md.; Architects: *Peterson & Brickbauer/Emery Roth & Sons*; Engineer: *Thornton-Tomasetti Engineers*. (Photo: © *Peter Aaron/Esto.*)

CONCLUSION

While concrete and steel are today's primary structural material for large projects, others have a definite place. Wood offers natural beauty, the ability to be constructed using hand tools, and good economy for some applications. Aluminum can be extruded in unique shapes and performs with minimal maintenance. Fiberglass and other fiber-reinforced plastics offer high strength, low weight, and corrosion resistance. Glass offers drama. Although high in cost, nontraditional materials will prove suitable for certain applications. As buildings, loads, and needs change, new materials will surely be developed.

7   *Parking Structures*

INTRODUCTION

What building type costs $8,000 to $10,000 per unit, rarely provides a good economic return, is regarded as a necessary evil, and, if not maintained, may last only 10 to 15 years? The answer, as building developers and owners are well aware, is the parking structure. The post–World War II explosion of family formation, suburban development, and private car ownership in the United States has led to the construction of thousands of parking structures with millions of parking spaces. Whether beneath an apartment house, within an office tower, or free-standing beside a shopping center, parking structures share several characteristics. They are exposed to public view (from inside and out), exposed to weather and temperature effects, and, in many areas, exposed to deicing salts or salty air.

SPECIAL CONCERNS

What are the concerns specific to parking structures? They include the following.

Initial Cost. Structure represents the lion's share of this building type's cost, especially where codes permit deletion of ventilation and sprinkler systems for open parking decks. Thus the structural design is under particularly close budget scrutiny.

Maintenance Cost. Structural slabs and joints must resist abuse from vehicular traffic and snowplows. Drainage lines must accommodate tracked-in sand and grit. Replacement of wearing surfaces must be considered.

Overall Temperature Changes. For most other structures climate-controlled interior spaces eliminate thermal movement concerns except at exposed elements and during construction. Parking structures experience daily and seasonal temperature swings throughout their lives. Note that the frequent air changes required to ventilate interior parking areas usually lead to some temperature variation there also.

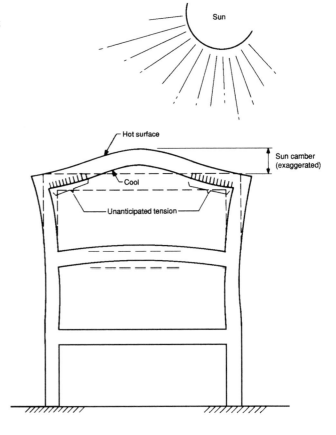

Figure 7.1 Sun camber occurs when the top surface of a deck is warmer than the soffit or supporting beam, causing a hump, or camber. This causes unanticipated forces and stresses in the supporting beam and column.

Differential Temperatures. The top level of an open parking deck will experience higher daytime temperatures (by solar gain) and lower nighttime temperatures (by radiant cooling) than levels below it. In addition, the variation in temperature through the depth of the floor beams can cause the top level to hump, or experience sun camber (Fig. 7.1).

Nonparallel Floors. The ramping inherent in most parking areas can provide ramp truss action for lateral loads. It can also create columns of varying heights, stiffnesses, and internal forces (Fig. 7.2).

Corrosive Attack. The use of deicing salts in cold climates has skyrocketed in recent years as the driving public has come to expect clear pavement, regardless of season. The salts are carried into parking areas on car underbodies and dropped there. Also, development in warmer climates has proceeded in oceanfront areas

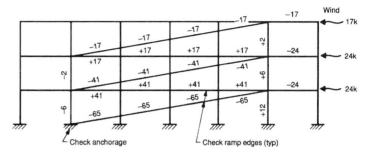

Figure 7.2 Wind forces accumulate in floor and ramp slabs in this example of ramp truss action. Such truss action can be helpful when specifically designed to carry lateral forces to the ground. It can be harmful if it causes unanticipated restraint against volumetric changes. Truss action cannot occur if the ramp is crossed by an expansion joint.

where salt is present in the air itself. While concrete normally protects embedded steel, chloride ions from salts negate this protection, permitting destructive rusting of steel and damage to concrete through the wedging action of the rust. Corrosion begins at chloride contents of 0.18 to 0.26 percent of cement by weight, equivalent to 1.2 to 1.5 lb/yd^3 of concrete, or 300 to 400 parts per million of concrete by weight.

Drainage. Most buildings require drainage at the roof, and appropriate pitch can be provided with minimal structural involvement. Parking structures require drainage at all levels, and the nature of construction usually dictates that the drainage slopes be integrated with the structural system for maximum efficiency.

Leakage. Drips from the floor above can damage a car's finish. Many approaches can be used to minimize leakage, including good drainage, crack reduction, joint sealing, and surface treatments.

Maintainability and Replaceability. Since some repair work is likely in the future, this aspect should be considered in design.

HISTORICAL PERSPECTIVE

It is interesting to note that many of these concerns did not become obvious until recently. For example, the model concrete code of the American Concrete Institute, ACI 318, considered garages as "moderate exposure" up through the 1977 edition. With the 1983 edition they were considered as "severe exposure" with recommendations for increased cement content or increased cover over rebar. In the 1989 edition cover recommendations were increased yet again.

The reason for this delayed recognition is simple. It took 15 to 30 years for the garages built in the 1950s and 1960s to deteriorate to the point that the problems became impossible to ignore. For example, a Minneapolis facility collapsed at 25 years of age, crushing numerous cars, but fortunately injuring no one. Although distress must have been visible earlier, the owner expressed his opinion that the structure "just got tired one day." In addition, maintenance practices are unsophisticated. When investigating a deteriorated parking structure, try talking to the custodian. He may proudly state that he uses no salt. But the drum of deicer he shows you may well have the active ingredient NaCl—salt!

Given current knowledge of parking structure concerns, what is the ideal solution? Fortunately a great deal of thought, research, and development has been brought to bear on the concerns. Unfortunately there is no ideal answer. As with life in general, the design of parking structures involves numerous tradeoffs and compromises. The approach finally selected for a particular project should have been reached with the full understanding of the owner regarding initial cost and maintenance requirements.

MAINTENANCE REQUIREMENTS

It is now good practice for the designer to provide a maintenance manual to an owner upon completion of the construction phase of the parking deck. The manual should include the following recommendations.

1. Use proper wash-down techniques (high-pressure spray from hose bibbs included in the design) to flush salty sediment away.

2. Use nonchloride ice treatments such as urea and sand.

3. Inspect expansion joints and sealants immediately before and after the winter. Repair any damage found.

4. Clean and inspect drainage systems, including floor drain sediment buckets, before and after the winter. Repair any leaks.

5. Inspect decks for cracks annually. Repair cracks by epoxy injection.

6. Periodically reapply sealing compounds (if applicable) on deck surfaces at intervals based on recommendations by the manufacturer or by observed wear patterns at high-wear areas such as aisles and turns.

7. Periodically inspect and repair waterproofing membranes (if applicable). Note that even a small tear admits water which can lead to progressive peeling of a membrane, so deferral of membrane repairs is not cost-effective.

GENERAL DESIGN RECOMMENDATIONS

Assuming that a good maintenance program will be followed, the following are some broad guidelines for improved parking deck durability.

Minimize Concrete Cracking

While chlorides can penetrate "solid" concrete through pores, a crack provides a much faster and more direct route; so, cracking should be avoided. In addition, water trapped in a crack can wedge it wider by freeze-thaw action.

Conventionally reinforced concrete is designed on the premise that concrete in tension will crack. Reinforcement is provided to resist the tension, providing needed strength. In addition, minimum reinforcement requirements, temperature-steel requirements, and the Z factor are code-specified design methods used to ensure that those cracks which do form will be small and well distributed. This is fine for normal structures subjected to minimal exposure to the elements, but is less than ideal under saltwater exposure.

While a design to eliminate cracking by eliminating top-surface tension sounds feasible (in a simple-span beam system, for example), it does not address shrinkage cracking. Concrete at the surface dries and shrinks before that deeper down. Therefore, the surface is in tension and the interior in compression even before external loads are applied. Shrinkage upon drying occurs because conventional concrete has excess water of convenience to give good workability. Some of this excess water rises to the top of fresh concrete as bleedwater. It must be allowed to evaporate before finishing operations begin, as forcing the water back in would create a weak surface. However, excessively fast evaporation can lead to plastic shrinkage, akin to the cracked mud in a dry lake bed. Under hot sun and dry winds, protection in the form of shading, foggers, and windbreaks may be required.

Even after the concrete has taken its initial set, cracking is possible if shrinkage stresses exceed concrete tensile strength. Retaining water during the cure gives the concrete a chance to gain strength before shrinkage stresses are imposed. For this reason a "water cure" is preferred, which can be provided by continually dampened burlap or by intentionally created ponds. Curing under an insulating cover which blocks loss of moisture is an acceptable alternative in freezing conditions where water use is impractical or unsafe. The thin films created by sprayed-on curing compounds may not give equivalent results under the open exposure and heavy foot traffic typical of parking deck construction.

In addition to good curing practice, an even more positive approach to crack control is to physically squeeze the concrete, eliminating crack-causing tension or, at least, squeezing closed any cracks which may form. This is done through prestressing (casting members with compression locked in) or posttensioning (compressing after concrete hardens by use of embedded tendons). The crack-reducing aspect of prestressing and posttensioning makes these approaches popular in parking garage construction, as discussed later. Where deicing salts are a concern, we recommend that the design strive to achieve a no-tension condition at the slab surface under dead load plus realistic live load of perhaps 1.2 kPa (25 lb/ft^2).

Another approach to reducing cracking in properly cured concrete is to include polypropylene fibers in the concrete mix. The fibers improve tensile strength, reduce the size of cracks, and distribute them evenly. However, their final effectiveness is still being studied.

Provide Good Drainage

Even in a "crack-free" surface, chloride-bearing water can penetrate concrete if allowed to stand long enough. Local depressions, or "birdbaths," are inevitably sources of trouble. All parking levels should be pitched to drains, since salty water can be carried to all levels by slush on car underbodies. In addition, all levels should be easy to wash down to remove salty sediments and improve cleanliness. Although roofs have been built with 1 percent pitch, a more generous 2 percent pitch 1/4 in/ft) is recommended for parking structures so that some pitch remains after normal construction tolerances, deflections, and cambers. Setting floor drains about ½ in lower than theoretically required will also aid floor finishers to avoid the common problem of birdbaths near the drain. In drainage layouts with two-way pitch, recognize that corners where two "high point" ridges meet can end up level unless one ridge is pitched also (Fig. 7.3).

Protect Against Freeze-Thaw Damage

An effective and economical way to avoid freeze-thaw damage is air entrainment. An admixture in the concrete mix can create microscopic air bubbles, with an air content in the fresh concrete of 4 to 6

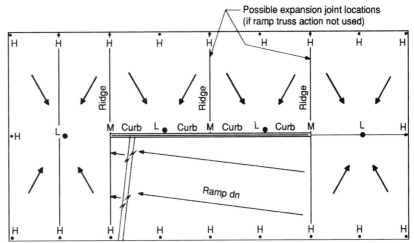

Figure 7.3 Sample drainage plan. •—column; H—high point or line; M—medium elevation (local high point); L—low point (drain); →—slab slope.

percent by volume. The bubbles provide "relief valves" which act when water in pores freezes and starts to expand. Air entrainment should be provided for all concrete members which may be exposed to freezing temperatures.

Minimize Concrete Permeability

The pores in concrete are a result of water of convenience. For the chemical process which gives portland cement its strength to fully occur, a weight of water equal to 20 percent of the weight of cement is needed (w/c=0.20). However, the resulting mix would be very stiff and unworkable, like dry bread dough. Additional water is needed to make the concrete workable, and this is termed water of convenience. Since the additional water is not chemically bound to the cement, most of it will eventually evaporate, leaving empty pockets and channels in the concrete.

The first strategy to minimize permeability is to keep the w/c ratio as low as possible, preferably 0.40 or less. This can be done by adding cement, but cement is more expensive than the sand it would replace, and increasing the volume of cement paste is likely to increase shrinkage. A more usual approach is to add a high range water reducer or "superplasticizer" to the mix. This admixture provides good workability with relatively little water of convenience. The low w/c mix also provides high concrete strength and rapid strength gain—a significant asset, since rapid construction and minimizing construction time are important economic advantages.

Once the concrete has been placed, cured, and allowed to dry, a second strategy is to fill, plug, or block the pores. Penetrating sealers are designed to soak into the top fraction of an inch of the concrete, where they are subjected to much less wear than at the surface. Some penetrating sealers are chemically reactive. They react with calcium hydroxide in the cement to form crystals which can fill pores, voids, and fine cracks. Silanes and siloxanes bond to concrete to form a water-repellent barrier around individual particles. Other penetrating sealers such as resins and methacrylates physically fill pores with inert material. Film-forming sealers block pores on the surface. Films are subject to wear, but are easily replaced by recoating.

Another strategy is to include a permeability-reducing admixture in the original concrete mix. Silica fume or microsilica is an ultrafine powder, a by-product of electronics manufacturing. When added to the concrete mix, the microsilica particles serve to fill spaces between other, larger particles and to chemically react with cement constituents. This results in reduced permeability and increased strength. In this case permeability is so low that the concrete will not absorb much, so follow-up sealers may become less critical. However, microsilica adds significant cost to a concrete mix, and requires special handling, finishing, and curing methods. The fresh mix is "sticky." It is also so impermeable that bleedwater does not rise. Therefore, an aggressive wet cure is required to avoid shrinkage cracking of the quickly drying water-deprived surface.

Protect Embedded Steel

If the planned structure may eventually experience chloride infiltration (as where no waterproof membrane is provided or maintained), consideration should be given to direct protection of steel. A time-honored, reasonable, and economical approach is to increase the concrete cover over the rebar. ACI concrete cover recommendations have grown over the years, from 16 to 19 mm ($\frac{5}{8}$ to $\frac{3}{4}$ in) to 25 mm (1 in) to 38 mm (1 $\frac{1}{2}$ in) and now 50 mm (2 in) at traffic surfaces of cast-in-place construction. While an extra fraction of an inch is helpful, providing too much cover can lead to poor control of surface cracking.

Another method of protection is to include a calcium nitrate admixture in the concrete mix for corrosion control. The admixture greatly reduces the rate of corrosion if and when chlorides reach the steel. This approach increases the cost of the as-delivered concrete.

A third approach is to use epoxy coatings. Epoxy-coated reinforcing bars are widely available. For the coating to have any value in improving structural longevity, the bars must be handled carefully. Any coating damage from cutting and bending must be touched up before installation, and epoxy-coated or all-plastic lathing tie wire must be used to avoid cutting the epoxy. Bars should be procured from certified plants for best results. Lap lengths must be increased per code, and ample cover must be provided where a fire rating is required. Epoxy coating is a very visible item in a construction budget, but should provide greatly improved durability.

For posttensioned structures, epoxy-coated tendons are now becoming available. Whether coated or not, tendons should be protected by maintaining their covers of corrosion-inhibiting grease and vinyl sheaths as intact as possible. Tape all cuts and gaps in each sheath, from anchor to anchor. To complete the protection, systems to encapsulate the end anchorages are available. The wires used in precasting are not amenable to epoxy coating or encapsulation, but as they generally occur at the bottom of members, they are farther from intruding salts and less subject to chloride attack.

Embedded items, which are common at precast connections, should be protected by hot-dip galvanizing or epoxy coating. Note that for items to be galvanized, recommendations of ASTM A143 should be followed. Cold-bent curves should be of generous radius (at least 3 and preferably 6 times plate thickness), punched holes and flame-cut notches should be avoided in thick material (over 16 to 19 mm, or ($\frac{5}{8}$ to $\frac{3}{4}$ in), and fluxless welding should be used if possible. Epoxy coating avoids the extra handling associated with hot-dipped galvanizing, but galvanizing will protect steel even if scratched, and damage to galvanizing can be touched up with zinc-rich paint.

Consider Using Waterproof Membranes

The most positive way to minimize leakage and avoid chloride damage is to keep salty water from ever reaching the concrete. A variety of practical traffic-bearing membrane systems are available, such as Kelman, Neogard, and 3M. In all cases, careful attention to substrate preparation and membrane installation procedures is required to attain satisfactory results. The membranes depend on the concrete floor slab for support, and if a membrane is placed on a dirty, greasy, or raveling slab, it will slip, tear, and lift. Note that a good-quality membrane, properly installed, represents a significant cost item. For this reason membranes are often resisted by owners, who understandably always have their eye on the construction budget. However, in many cases an owner is willing to install a membrane after years of operation in an attempt to halt the progression of chloride damage. Had the same membrane been applied initially, it could have been provided more economically, adhered more securely, and resulted in a longer lasting structure.

Note also that a membrane is not a cure-all. Good drainage, housekeeping, and maintenance are still required. Tears and worn spots must be repaired promptly before moisture can work under the membrane and cause progressive loss of adhesion.

Anticipate and Provide for Volumetric Changes

Volumetric changes include creep, shrinkage, thermal movement, and sun cambering. Creep is the slow, gradually decreasing cumulative shortening which is exhibited by concrete under sustained load. For high-rise buildings, vertical creep of columns is significant (see Fig. 4.5). For parking structures, horizontal creep under prestressing or posttensioning forces must be considered. In effect, slabs and beams try to shorten, but columns and walls which are anchored to the ground try to resist this movement.

Shrinkage, as discussed, is the shortening of members over time as free water of convenience evaporates. This will begin when curing is stopped and will end when member moisture content is in equilibrium with ambient relative humidity. For thick members and dry climates this can take months to years.

Thermal movement can be lengthening, shortening, or bending of members, depending on the direction and gradient of temperature change.

The forces induced by volumetric changes can be so large that designing for sufficient strength to resist them is impractical. In certain cases, such as prestressed and posttensioned members, resisting the forces would also be counterproductive, since the desirable crack-closing compressive forces would be diverted into the supports and away from the member of concern. Strategies to deal with volumetric changes include the following:

1. Locating lateral load-resisting elements such as stiff columns, shear walls, and elevator shafts toward the center of building segments if possible (Fig. 7.4a).
2. Orienting lateral load-resisting elements to provide minimal restraint against movement (see Fig. 7.4c).
3. Providing expansion joints, typically not more than 60 m (200 ft)

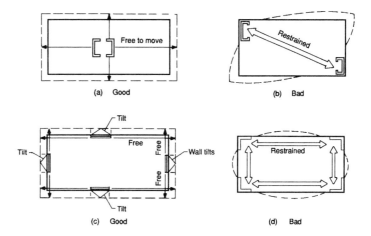

Figure 7.4 These plans show approaches to allow for volumetric changes (shrinkage, creep, and thermal movement) in slabs. (*a*) Good. Central core permits local slab movement. (*b*) Bad. Opposing cores lock in slab restraint. (*c*) Good. Walls centered on faces can tilt for movement; slab is anchored laterally by walls but has local freedom of movement. (*d*) Bad. Walls located at corners cannot tilt; slab restraint is locked in by opposing walls.

apart or 30 m (100 ft) from a restraint. Due to inherent "give" in connections, the Prestressed Concrete Institute (PCI) recommends joint spacings of 90 m (300 ft) and 45 m(150 ft), respectively, for precast structures if not in the primary spanning direction.

4. Providing double columns at expansion joints, if feasible. Low-friction slip pads such as Teflon PTFE can also be used on brackets from single columns, if properly detailed. Neoprene pads are not a direct substitute for PTFE. Neoprene requires at least 2.8 MPa (400 lb/in^2) bearing to avoid "walking," and ample thickness to accommodate movement by shearing deformation—at least twice as much thickness as anticipated movement from the "neutral" position.

5. Reducing shrinkage by curing as long as practical, and by steam curing (precast elements only).

6. Reducing creep by letting concrete gain as much strength and stiffness as possible before loading. Using high-strength, hard-

aggregate concrete acting at low stresses is also recommended.

7. Reducing creep by adding rebar to compressive regions of the concrete.

8. Allowing as much creep and shrinkage as possible to occur before engaging restraining elements. For precast this means using members at least 28 days old where practical. For cast-in-place construction this means casting and stressing in stages (so later stages are not affected by the movement which already occurred in earlier stages) or using pour strips (so small areas can move independently before they are tied together).

9. Avoiding connections from retaining walls to precast or post-tensioned structures which resist volumetric changes. Ideally this means designing retaining walls to be free-standing. This also simplifies construction sequencing. However, successful parking decks have been built where the slabs brace the walls and relative movement is permitted by plastic slip strips at the interface. Such walls must be designed anticipating later movement of the supporting slab (Fig. 7.5). Nonprestressed decks have used pour strips between slab and wall, with pour strip concrete placed as late as possible to allow most creep and shrinkage to occur. Rigidly locking the deck to the wall is not recommended for prestressed and posttensioned structures, regardless of timing.

10. Designing reinforcing (if monolithic) or connections (if precast) to handle the stress reversals or movements which can occur due to sun camber at top-level framing (see Fig. 7.1) and due to overall thermal movement at the levels just below the top and just above grade.

11. Providing control joints to direct cracking. In any structure some cracking is inevitable. It is better to provide neat, defined, sealable joints than to have random cracks. In this way leakage paths are minimized and corrosion protection is improved. Provide control joints, tooled to receive sealant, at:

- All construction joints in slabs, to block these "built-in" cracks
- All girder lines in steel structures, where beam deflections "kink" the slab
- All joints between precast floor members (in the topping), as differential movement inevitably occurs and initiates cracking
- Posttensioned slabs, at column lines where rebar shrinkage or

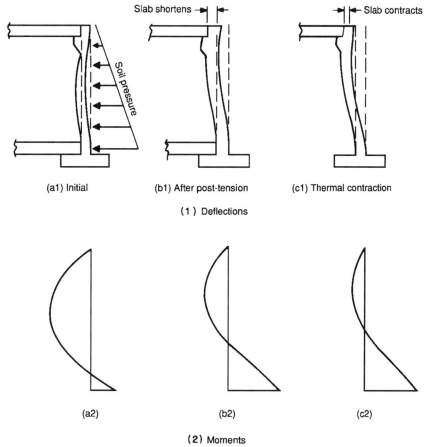

(1) Deflections

(2) Moments

Figure 7.5 When supporting a foundation wall by a posttensioned slab, conditions vary over time. (1) Deflections. (2) Moments. (a1, a2) Upon backfilling, wall acts primarily simple-span. (b1, b2) Slab shortens when tendons are pulled, and base fixity provides some cantilever action. (c1, c2) Subsequent thermal movement may change moments again.

temperature reinforcing is used, as in small, odd areas impractical to "squeeze" with posttensioning

- Rustication joints on spandrels, adjacent to columns

Design Expansion Joints and Joint Covers with Care

Expansion joints are a necessary nuisance on long parking decks. The designer must consider their function, behavior, and exposure. First, expansion joints should always be located along high points of the floor slab (see Fig. 7.3). Second, the anticipated range of movement should be established conservatively. We recommend

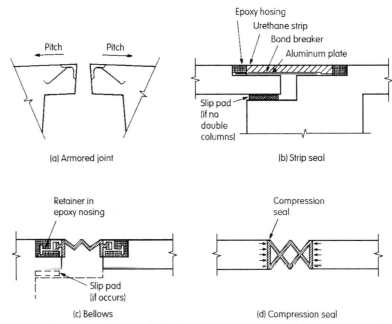

Figure 7.6 These sample expansion joints are sized to accommodate similar amounts of movement. (*a*) Armored joint. Open, flared shape is not watertight, but also is not subject to damage or wedging by debris. (*b*) Strip seal. Proprietary systems use an elastomeric strip to accommodate movements at lightly loaded joints. This joint will not trap debris, but can be damaged. (*c*) Bellows. Mechanically fastened seal has labor-intensive installation. (*d*) Compression seal. System is easily installed, but performance is sensitive to quality of preparation.

that creep and shrinkage estimates be rounded up, as it is not uncommon for a joint cover to be torn by unexpectedly large movements. Slip pad bearing areas and travel lengths should also be generous. Third, construction documents should show the design joint width for a specific temperature and the adjustment in width required if the temperature is different at the time of construction. Fourth, the designer and owner should consider whether an expansion joint cover is really necessary. A simple armored-edge open joint could be kept very narrow, making it less subject to damage. By flaring the joint wider below the surface, debris will drop through and not get trapped (Fig. 7.6*a*).

The perfect expansion joint cover would be inexpensive, durable, and easy to repair. It does not yet exist. The heavy-duty joints used on highways are unnecessarily costly for parking decks. Elastomeric strip seals are often used instead. They have a factory-molded urethane rubber strip, supported from below by an aluminum plate which bridges the joint gap. The strip is fastened to the deck on each side by an epoxy nosing (Fig. 7.6*b*). If the strip is ordered too narrow, crosswise tensile stresses will tear it when the joint opens. Since the strips are generally 20 to 30 cm (8 to 12 in) wide, they can be snagged and gouged by snowplows, ripped by the shearing stresses of stationary cars turning their wheels, and pounded by heavy traffic (if bus or truck traffic is permitted to cross). But the strips offer economy, the ability to follow irregular joints, and the ability to be repaired easily. Being flush, they also will not trap debris.

Bellows-type joint covers, anchored to each slab edge with metal nosings set in epoxy, offer a much narrower exposed face than elastomeric covers and so are less susceptible to damage. However, debris can be trapped in the joint. Also, splicing the bellows and following jogs in the joint can be difficult (Fig. 7.6*c*).

Compression seals work in theory and lab tests but are difficult to install properly. Smooth, square, clean, parallel, even joint sides and careful placement are needed to result in a watertight seal which does not "walk out." This is a challenge in the real world of construction under tight budgets and schedules (Fig. 7.6*d*).

Consider Exceptional Loadings

While the common code-mandated live load of 2.4 kPa (50 lb/ft^2) is known to be much greater than passenger cars can apply, decks should also be checked for capacity under emergency vehicle access where applicable (if a firefighter *can* drive there, he or she *will* drive there!), accidental truck loading (particularly where the deck is flush with surrounding grade and does not look like a garage), and snowplow and snow pile loading, if applicable. Snow removal should be worked out with the owner during design, as handling systems can be quite elaborate in snowbelt areas.

OVERVIEW OF STRUCTURAL SYSTEMS

There is no single "best" system for parking structures, as the selection process must consider parking layout, structure above (if any), availability of materials and labor, foundation conditions, exposure, protection systems, and tradeoffs of first cost, maintenance cost, and desired minimum useful life.

The overview in Table 7.1 is organized in order of increasing span and increasing weight. Relative costs are not addressed as they vary by location. Schematic diagrams keyed to the table are shown in Figs. 7.7 to 7.13.

Table 7.1 Overview of Structural Systems

Type	Advantages	Disadvantages
Short-span steel with slab on metal deck (Fig. 7.7)	Light weight Fast erection (no formwork) Works with steel frame above (if occurs) Easy to modify and perform remedial work Easy to inspect	Columns create inefficient parking layout Slab must include rebar (deck may rust out) Not watertight—waterproof membrane strongly recommended Deeper construction than flat plate, flat slab Repainting eventually necessary
Short-span concrete flat plate and flat slab (Fig. 7.8); waffle and` joist systems similiar	Conventional concrete construction Works with concrete (or steel) frame above Simple formwork Shallowest floor-to-floor height (waffle/joists deeper) Modifications and remedial work feasible on a bay-by-bay basis No painting required	Inefficient parking layout Needs formwork, reshores Not watertight— waterproof membrane strongly recommended
Long-span steel with composite joists and slab on metal deck (Fig. 7.9)	Efficient parking layout Light weight Fast erection (no formwork) Easy to inspect	Slab is thin and not watertight —waterproof membrane strongly recommended Repainting joists is difficult Specialized joists required
Long-span steel with posttensioned slab (Fig. 7.10)	Efficient parking layout Fast erection No conventional shored formwork—steel-supported reusable forms are used Posttensioning tends to close cracks (may be one-way or two-way), improve watertightness Lighter than equivalent posttensioned concrete Can route services through girders	Specialized posttensioning trade involved Specialized fabrication (if castellated) Specialized erection (strongback may be required) Repainting eventually necessary Remedial slab work difficult
Long-span precast concrete (Fig. 7.11)	Efficient parking layout Fast erection Good quality control and rebar protection No painting required	Specialized erection required —large, heavy pieces Heavy overall structural weight Installation and maintenance of numerous joint seals (between each piece)

Table 7.1 Overview of Structural Systems (cont.)

Long-span cast-in-place posttensioned concrete (Fig. 7.12)	Efficient parking layout Two-way posttensioning gives tight slab, minimal joints, good rebar protection (sealer recommended) No painting required	Specialized posttensioning trade required Labor-intensive construction with formwork Construction weather-sensitive Overall structural weight Remedial work difficult
Precast formwork with cast-in-place topping (Fig. 7.13)	Efficient parking layout (in long-span system) Finished bottom without formwork No posttensioning trade (precast may be prestressed) No painting required	Precast requires shoring Specialized precast required Labor-intensive rebar and topping placement Slab is not watertight —water proof membrane strongly recommended Overall structural weight.

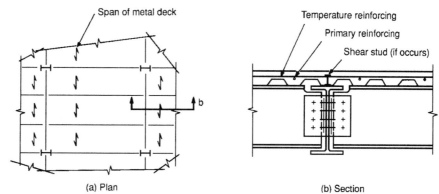

(a) Plan

(b) Section

Figure 7.7 Short-span steel-framed deck. (*a*) Part plan. (*b*) Section.

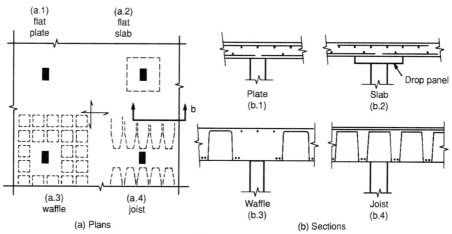

(a) Plans

(b) Sections

Figure 7.8 Short-span concrete deck options. (*a*) Plans. (*b*) Sections. (*a*1, *b*1) Flat plate. (*a*2, *b*2) Flat slab (drop panel). (*a*3, *b*3) Waffle slab (two-way ribbed). (*a*4, *b*4) Joist (one-way ribbed).

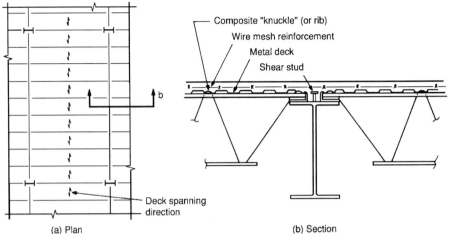

(a) Plan   (b) Section

Figure 7.9   Long-span steel framing with metal deck. (*a*) Plan. (*b*) Section. Reinforcing mesh, not metal deck, provides the tensile strength needed.

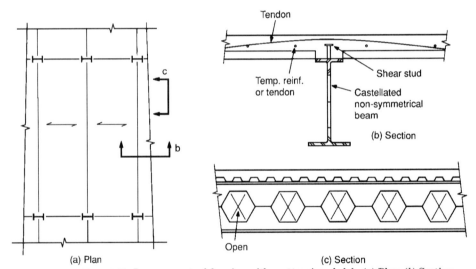

(a) Plan   (c) Section

Figure 7.10   Long-span steel framing with posttensioned slab. (*a*) Plan. (*b*) Section through steel girder. Note draped tendons. (*c*) Elevation of steel girder. Castellated girder is shown for lighter member weight and through-web access.

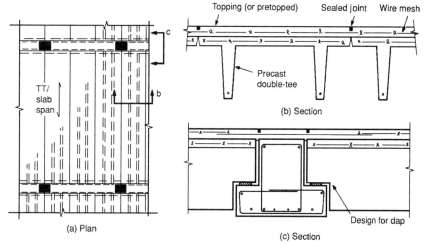

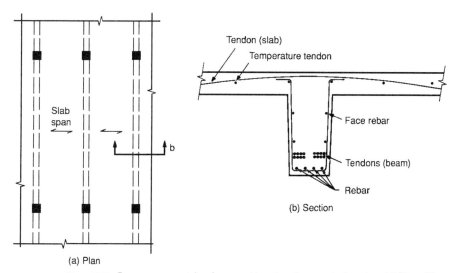

Figure 7.11 Long-span precast concrete framing: (*a*) Plan. (*b*) Section through long-span double-T deck members. Sealed joints are recommended over T joints. (*c*) Section through short-span inverted-T girder. Sealed joints and checks of beams at daps are recommended.

Figure 7.12 Long-span, cast-in-place, posttensioned concrete framing. (*a*) Plan. (*b*) Section through long-span floor beam.

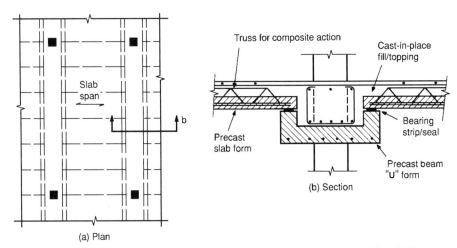

Figure 7.13 Precast forming system with cast-in-place topping. (*a*) Plan. (*b*) Section through long-span floor beam/U-form.

CASE HISTORIES

Because durability is the major concern in parking decks, these case histories will highlight a variety of problems that have been encountered. Inclusion or exclusion of a particular structural type should not be interpreted as condemnation or approval of that type, as all types can perform well or poorly depending on design, construction, and maintenance.

Case 1: Steel Framing with Slab on Metal Deck

Metal-deck parking structures (see Fig. 7.7) have required extensive reconstruction due to three problems. First, some were constructed with level floors, so salty water was retained in puddles. This was particularly common within buildings, where drainage needs were not obvious. Second, metal deck tends to trap water against it. This is a challenge for any coating to resist. As discussed in Chap. 5, even weathering steel does not form a tight, self-limiting oxide coating under conditions of continual moisture. Third, the metal deck used was of a composite type and was designed to act as the slab reinforcing.

On these "flat" floors, puddled water penetrated the slab at midspan and was retained on the metal deck, rusting it out. Therefore, the location requiring the most bottom steel actually ended up with the least. At such structures, demolition and reconstruction of the slabs was required.

It is interesting to note that, while this system can have significant durability problems, it also offers an important advantage during reconstruction. Due to its simple span nature, demolition and replacement can be limited to slabs alone and can proceed on a bay-by-bay basis without danger of instability or progressive failure.

Case 2: Concrete Flat Plate

Typical of numerous residential parking garages, this deck consisted of below-grade short-span slabs (see Fig. 7.8) with minimal pitch. Deterioration was of two types. First, top rebar at columns had rusted, delaminating the slab concrete around each column (Fig. 7.14). This was due to a combination of inadequate rebar cover in the original construction (generally 1 in or less), the negative-moment flexural cracks which occur naturally around columns and provide easy entry for salty water, and the lack of pitch away from columns. Second, slab soffits (the ceilings of floors below) had spalled, exposing severely rusted bottom rebar (Fig. 7.15). Here water had

Figure 7.14 Flat plate slab discussed in Case 2. Note damage around column and minimal cover for top layer of rebar. (Photo: *Charles Thornton.*)

Parking Structures

Figure 7.15 Flat plate soffit discussed in Case 2. Note damage at midspan and loss of upper layer rebar area. (Photo: *Charles Thornton.*)

puddled and penetrated at natural low points of this "flat" floor.

Repairs consisted of removing delaminated and weakened concrete around columns, replacing it with patches of high-density sealed concrete which was humped to shed water and was thick enough to provide adequate cover. For soffit spalls, floor drains were added at these natural low points and patches were made locally or as through-slab concrete pours, with rebar added to replace and lap beyond rusted-out reinforcement areas.

Here again the nature of short-span construction permitted working on a bay-by-bay basis without adverse impact on the overall structure. Interestingly, the floor system operated many years without adequately bonded top steel. We believe this may be due to the 3.6 kPa (75lb/ft^2) unreduced live load required originally, compared to a 1.2 kPa (25 lb/ft^2) realistic actual live load.

Figure 7.16 Cantilevered joist framing discussed in Case 3. Typical framing is at top. Arrow shows crack along girder, passing to left of columns. (Photo: *Len Joseph.*)

Case 3: Long-Span Conventionally Reinforced System

A pan-joist system illustrates a disadvantage of long-span conventionally reinforced cast-in-place systems. Negative moments that are adequately reinforced for strength may still develop objectionable and troublesome cracking.

As shown in Fig. 7.16, the designer located a line of supports about one-fourth of the way into an 18-m (60-ft) double-loaded bay. Structurally this was advantageous to address a constrained foundation condition and to reduce long-span positive moments. However, a cursory inspection reveals a flexural crack in each floor directly aligned with the cantilever support girder. This crack must be sealed, maintained, and monitored to avoid its being a cause of premature deterioration.

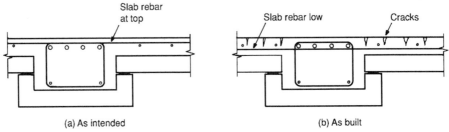

Figure 7.17 Misplaced rebar discussed in Case 4. (*a*) As intended, slab rebar would be in topmost layer. (*b*) As built, slab rebar was second layer.

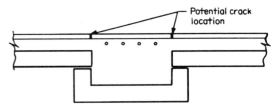

Figure 7.18 Potential crack location adjacent to beam, as discussed in Case 4.

Case 4: Composite Precast Formwork System

Thin precast, prestressed concrete planks form slab soffits and U forms create beam soffits in this system (Fig. 7.13). Loops projecting upward connect a conventionally reinforced topping layer to these forms. While positive moment (tension at bottom) uses the prestressing in the forms to advantage, negative moment (tension at top) can still cause significant cracking, so field control of rebar placement is still critical.

In one case, extensive and severe cracking occurred in a slab top surface soon after project completion. Investigation revealed that the field-placed top rebar was set in the wrong order. Slab steel was to be topmost, with beam steel in the second layer, but this order had been reversed (Fig. 7.17). In a normal slab system it would make little difference, but in this case placing the slab rebar below heavy beam rebar moved slab steel down to slab middepth, where it was virtually useless.

Even in a deck where construction is properly performed, the use of conventional rebar to resist negative moment can still result in slab cracks at the slab-to-beam interface. Such cracks must be sealed, maintained, and monitored (Fig. 7.18).

Case 5: Precast Systems

The crack-closing property of prestressing, combined with good-quality control on rebar placement, concrete mixes, and curing at

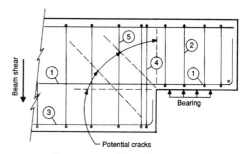

Figure 7.19 Items to consider when designing a beam dap (notch). (1) Develop both ends of dap tension rebar crossing potential crack zones through dap corner and overall beam. (2) Provide shear reinforcing at dap based on reduced member depth. (3) Develop main flexural rebar to take force from item 1 bars. (4) Provide additional stirrups sufficient to "hang" the full shear load. (5) Provide standard stirrups in main member away from the dap.

the precasting plant mean that problems in precast garages rarely occur in the body of the typical members themselves. Joints and connections are the likeliest locations for trouble.

Cracking at daps (notches at beam ends) and cutouts are common problems which reflect inadequate attention. The designer must call for additional reinforcing in these areas, and the precaster must check that these nontypical conditions are properly constructed (Fig. 7.19).

Leaking joints, a common problem in older decks, reflects a lack of maintenance. Owners must recognize that precast systems will have many more linear feet of sealant to maintain than monolithic systems. Even where the precast deck receives a cast-in-place topping, the possibility of reflective cracking makes it advisable to have a control joint cut into the topping over each precast joint and sealed.

Cases 6 to 9: Posttensioned Systems

The large forces provided by posttensioning tendons are a two-edged sword. On one side, the draped tendons physically lift the slab and beams to "balance" most of the dead load, while compressing the concrete to reduce tensile stresses and greatly reduce cracking. On the other side, the tendon forces are applied in small local areas, creating irregular stress concentrations. Also, when most of the load is balanced, small changes in the total load will result in large changes in unbalanced load. In addition, the monolithic

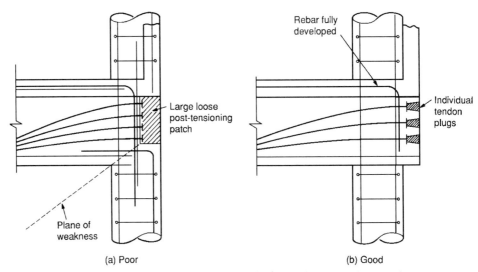

Figure 7.20 Anchorage details at round columns. (*a*) Poor. A large pocket creates a plane of weakness in the column and requires a large, easily dislodged patch. (*b*) Good. Extend the beam profile through the column. Conventional reinforcing and anchor details can be used.

nature of cast-in-place concrete and the long-span nature of posttensioned systems make frame action an important consideration.

Case 6: Anchorages and Movement

At one parking deck, beam-end anchorage details caused a problem when they conflicted with architectural appearance. Normally the anchorages are placed at the outside face of the supporting column, but with a round column there was not enough room on the curved face. The engineer agreed to set the anchorage at column middepth. This caused three problems:

1. The anchorage plates created a plane of weakness within the column.
2. The patch to cover the anchorage, representing half the column, was not secure.
3. Mild-steel rebar from beam to column had to be bent down too early, creating a weak beam-to-column joint.

The result was severe deterioration at every beam-to-column joint, requiring premature demolition of the deck (Figs. 7.20 and 7.21).

Another problem at this deck was differential movement. A masonry enclosure for a stair was rigidly attached to each level of the deck. As the levels moved differently due to solar heating, the

Figure 7.21 At the posttensioned deck of Case 6, vertical cracks at column centerlines show the detrimental effect of moving tendon anchor plates from their usual outside-face locations. (Photo: *Len Joseph.*)

masonry was damaged (Fig. 7.22). Nonstructural elements which bridge between structural elements should be isolated by soft joints.

Case 7: Missing Dead Load

If the nature of load balancing by posttensioning is not understood by the contractor, construction problems can occur. In one case, girders with many posttensioning cables were intended to pick up columns from a steel-framed office building above. However, the garage contractor wished to complete his work before the upper building was constructed. He tensioned the tendons, creating an upward lift, before there was dead load available to resist it. The girders cracked along a plane of weakness created by the numerous tendons placed within (Figs. 7.23 and 7.24). The lesson here is to consider and specify the sequence of loading and tensioning.

Case 8: Force Patterns during Construction

Due to the monolithic nature of concrete frames, force patterns can differ from assumptions. On a five-story garage, a second-

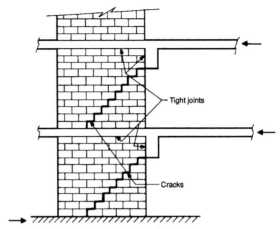

Figure 7.22 Stiff but nonstructural concrete block stair walls in Case 6 cracked as they tried to resist movement between two floors. Soft joints at wall tops and sides would avoid this damage.

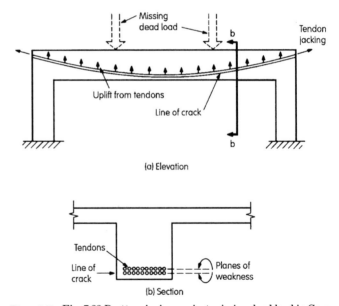

Figure 7.23 Fig. 7.23 Posttensioning against missing dead load in Case 7. (*a*) Elevation shows tendons providing a lifting force greater than the dead load currently present. (*b*) Section shows numerous tendons creating a plane of weakness across the beam.

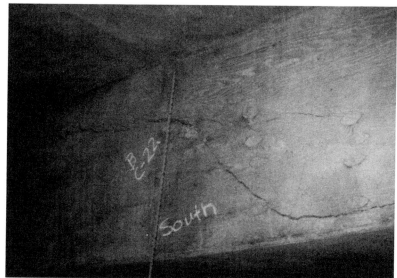

a

b

Figure 7.24   Pattern of beam cracking in Case 7. (*a*) Shear/flexure cracking near the end of one beam. (*b*) Cracking near midspan of another beam. (Photo: *Thornton-Tomasetti Engineers.*)

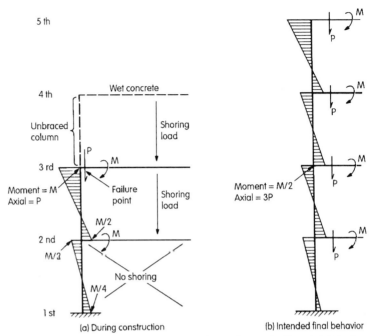

Figure 7.25 Force pattern explanation for column distress in Case 8. (*a*) During construction. Wet concrete at level 4 created a large moment but a small axial load at level 3. (*b*) Intended final behavior. Moment at level 3 was to be only half of that during construction, and helpful axial load was to be three times greater.

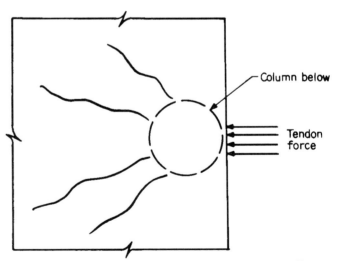

Figure 7.26 Plan of bursting stress cracks from tendon thrust in Case 9.

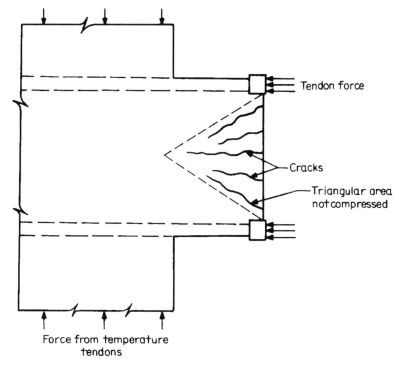

Tendon force

Cracks

Triangular area
not compressed

Force from temperature
tendons

Figure 7.27 Plan of secondary slab cracks in Case 9, as induced by the simultaneous spreading action of tendons loading beams at slab edges and lack of compression in the area between beams.

floor column was damaged due to unanticipated moment during construction.

While pouring concrete on level 4, shoring distributed the weight to levels 2 and 3. A simple review indicated that the wet concrete weight would not exceed the design live load. However, moments at the column just below level 3 were much greater in this configuration than in the completed structure, due to a lack of column restraint at level 4. In addition, column moment strength was reduced since full axial load was not acting on the column (Fig. 7.25). As a result a shear/moment hinge formed just below level 3. The lesson here is to avoid asking an incomplete structure to carry full design loads. If full shoring to grade is impractical, analyze the partial structure for construction-stage loadings.

Case 9: Secondary Effects from Tendons

Forces from posttensioning tendon anchorages can have subtle effects on slab stresses. In one case, lack of compression-block reinforcement permitted bursting stresses to develop, causing a pattern of radial hairline cracks in the slab above (Fig. 7.26). In another case, compression at two posttensioned beams caused cracking at the slab between them (Fig. 7.27). In both cases, recognition of the potential for cracking and addition of mild steel to limit the cracks would have been helpful.

CONCLUSION

Parking structures represent an area of design where economic tradeoffs are particularly significant. As tradeoffs among first costs, maintenance costs, and building life will influence the choice of structural system, full involvement of the building owner is needed in making this decision. The contract drawings, specifications, and other directions to the contractor must consider the particular characteristics of the system selected to avoid surprises. Upon building completion, the building owner should be informed of proper maintenance procedures to keep the structure in good shape. If this approach is followed, future parking structure problems will be minimized.

8

Long-Span Structures

INTRODUCTION

When you think of a "wow!" architectural space, the most likely image to come to mind is a long-span structure.

Whether arena, stadium, field house, or convention center, the action is all down on the field, floor, or track. Then why expose the structure? The first and most persuasive reason is money. Ceilings are customarily used to conceal the potentially messy tangle of ductwork, conduit, and lighting fixtures common to office spaces. In large, open arena spaces, ventilation distribution systems can be confined to perimeter mechanical rooms or isolated rooftop package units. Lighting can be a widely spaced pendant type, and since the light is directed down, the roof deck area is dark and less noticeable, so why spend money to conceal it?

Once a ceiling is deemed unnecessary, interaction between architect and engineer begins in earnest as discussions work toward finding a system appropriate to the space which is both visually pleasing and structurally efficient. Efficiency is normally an important consideration as the structure of a long-span space makes up a larger fraction of total building cost than the structure in more conventional spaces.

Efficiency should be carefully evaluated, as often the most economical structure is not the one with the fewest pounds of material, but the one which local fabricators and erectors best understand and are most comfortable with.

Long-span structures come in such a wide variety of types that this chapter will serve only as an overview of the kinds of systems available. Design of any particular system is beyond the scope of this book.

FLEXURAL SYSTEMS

These systems cover space like a beam, using top and bottom flanges or chords to resist tension or compression and webs or diagonals to resist shear. They are the most versatile systems as they

Figure 8.1 This trussed folded-plate roof has individual trusses for each plane. Pairs of trusses were married to form stable inverted-V units for easy erection of the 37-m (120-ft) span.
United Airlines Terminal, O'Hare Airport, Chicago, Ill.; Architect: *Murphy/Jahn with A. Epstein*; Engineer: *Thornton-Tomasetti Engineers.* (Photo: *Len Joseph.*)

can accommodate virtually any plan shape, boundary conditions, and roof profile, including steps (see Fig. 3.25). Special roof openings can also be accommodated if originally designed for. However, they have two drawbacks. First, they are not as structurally efficient as arches and domes (where those shapes are feasible for a particular application). Second, they are generally not self-supporting until fully assembled. This means that some type of falsework is required to hold individual pieces or sections during assembly. In a system with several larger pieces, a few falsework towers may suffice, while the numerous small pieces of most space truss systems may require a complete scaffolded floor. Self-stabilized trusses such as modules of folded-plate trusses can simplify erection while providing visual interest (Fig. 8.1).

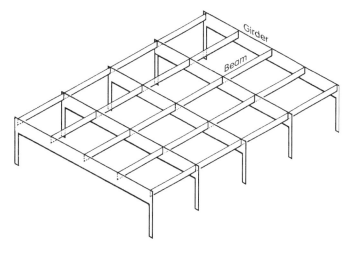

Figure 8.2 A basic framing system uses simple-span beams, girders, and columns.

Girder

Beam

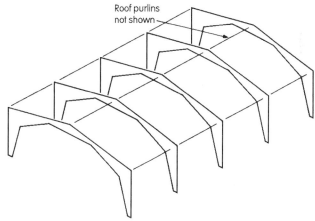

Figure 8.3 Rigid frame systems are most efficient when the frame shape can be fine-tuned to the applied load.

Roof purlins not shown

Simple-Span Beams and Girders

Simple-span beams and girders form the most basic flexural system, but for any appreciable span the member sizes and weights grow so large as to be unwieldy. Therefore, this system is normally used only for heavily loaded roofs, or where picking up floors above. In these applications the framing self-weight has less relative importance (Fig. 8 2).

One-Way Rigid Frames

One-way rigid frames form the next simplest system (Fig. 8.3). By rigidly connecting girders to legs, maximum moments can be

---

reduced and members can be made lighter. Preengineered building manufacturers have developed design programs that result in extremely efficient frames, if they are permitted to establish member proportions. An advantage is that frames can be assembled flat and then tilted up, avoiding falsework. A disadvantage is that rigid frame legs must occur at predetermined locations.

Propped Beams with Mast Columns

Propped beams with mast columns form a variation on the rigid frame (Figs. 8.4 and 8.5). Cables or rods running from beam over mast and down to grade can provide the advantage of a rigid frame's "knee," but with more visual interest. Of course, the special fittings and more intricate members would increase the cost.

Rigid frames can also be constructed in concrete. This makes sense only when heavy roof loads are anticipated so that the concrete frame's weight penalty becomes relatively less significant, as at the Moscone Convention Center in San Francisco.

Trusses

Trusses can be used for longer spans. One-way trusses, while not particularly eye-catching, are familiar to fabricators and erectors.

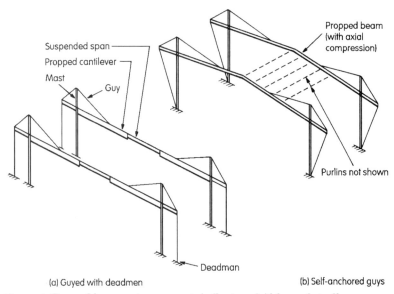

(a) Guyed with deadmen (b) Self-anchored guys

Figure 8.4 Propped-beam systems can act similar to a rigid frame, but offer more visual interest. (a) Guyed with deadmen. (b) Self-anchored guys.

Figure 8.5 The nature and function of this propped-beam system with deadman guys is clear.
Darling Harbour Exhibition Centre, Sydney, Australia; Architect: *Philip Cox, Richardson, Taylor*; Engineer: *Ove Arup & Partners.* (Photo: *Ove Arup & Partners.*)

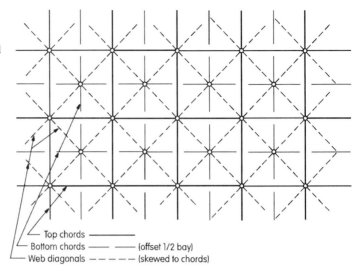

Figure 8.6 In this plan of a conventional space truss (space frame) system, top and bottom chords are parallel and offset one-half module. Webs are skewed.

Top chords ——————
Bottom chords —— —— (offset 1/2 bay)
Web diagonals — — — — (skewed to chords)

Therefore, bid prices are usually quite competitive. One-way trusses can be assembled on the ground, lifted, and placed without falsework. Two-way trusses, intersecting to form a grid like a tic-tac-toe board, can offer some reduction in steel tonnage by making use of supports on four sides of the spaces (where applicable) rather than two sides. However, fabrication and erection will be complicated by the truss intersections. Where desired, a two-way appearance can be economically provided by using one-way trusses with similarly patterned bridging lines in the perpendicular direction.

To simplify the truss intersection problem, manufacturers have developed proprietary joints and standardized member sizes. In this way a continuous top chord grid, continuous bottom chord grid, and continuous web of diagonals can be assembled efficiently. This forms a space truss, often called a space frame (Fig. 8.6). Web diagonals can either lie in a vertical plane (top and bottom chord grids align) or sloped to form inverted pyramids (grids offset one-half module). In another truss layout the top chord grid defines the base of the web pyramid, while the bottom chord grid is turned 45° to run on a skew between pyramid apexes. This way there are fewer bottom chord members to handle and, if the bottom chords are in tension, there is no unbraced length penalty (Fig. 8.7a). Skewed grids can also mediate between different building planes (Fig. 8.7b). Examples of proprietary space truss systems are shown in Fig. 8.8 a–d. Their suitability varies with desired span, depth, load, and

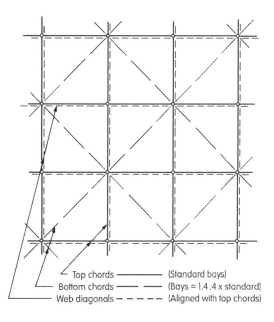

Figure 8.7a This space truss system plan has chords skewed 45° and webs aligned with top chords.

Top chords ———————— (Standard bays)
Bottom chords — — — (Bays = 1.4 .4 x standard)
Web diagonals — — — — (Aligned with top chords)

Figure 8.7b This space truss uses skewed top and bottom chords to reconcile differing roof and wall lines.
St. Francis Hospital, Waukegan, Ill.; Architect: *American Building Systems.* (Photo: *courtesy of Mero Structures.*)

Figure 8.8 Manufacturers offer a wide variety of space truss systems, including this small sample. (*a*) Unistrut System III ball node. (Photo: *Unistrut.*). (*b*) Unistrut System II box node. (Photo: *Unistrut.*) (*c*) Unistrut System VI (shorter span). (Photo: *Unistrut.*). (*d*) Multihinge (nodeless) connection system of Pearce Structures. (Photo: *Peter Pearce, Pearce Structures.*)

geometry. (Some can even accommodate tapered grids and depths.)

As a space truss depends on two-way action, falsework is required to support the incomplete structure during erection. Erection may proceed member by member on raised staging, member by member on the ground and then jacked into position, or by lifting preassembled modules (although this is most susceptible to fit-up difficulties).

When originally developed, the space truss was touted as being safer than other structural systems due to its redundancy. The survival of war-damaged space truss roofs supported this view. However, recent reliability studies and investigations of such failures as the Hartford Civic Center Coliseum highlight the fact that while redundant structures which are properly designed and constructed have reserve capacity, conversely any error or weakness in design or construction of a repetitive element can lead to a "zipper," or progressive, type of failure. Therefore, space trusses require as much attention to design and construction as other systems, if not more so.

Eight rules of thumb are offered by space frame designer and manufacturer Peter Pearce of Pearce Structures, Inc.:

1. Frequent, symmetrically located supports inboard of the roof edge improve efficiency. An overhang of about one-third the interior span helps to counterbalance the interior span.

2. Efficient span-to-depth ratios for simple-span conditions are 12:1 for one-way spans, 15:1 for two-way spans, and 18:1 for three-way (triangular) spans.

3. The depth-to-grid ratio should be between 0.5:1 and 1:1. Where greater depth is required, going to a triple-layer system (top, middle, and bottom chords) may make sense. Member lengths and grid modules can then be kept within standard ranges, and the extra layer can be eliminated where not needed, resulting in a variable-depth space frame.

4. Grid sizes should be between 1.2 and 3.6 m (4 and 12 ft). An efficient range for standard modular systems is 1.5 to 2.1 m (5 to 7 ft), subject to the module of the elements being supported—skylights, roof deck, and so forth.

5. Use configurations generated from a uniform module or unit cell such as a square, rectangular, triangular, or radial grid. Some systems accommodate variable modules, but this is less efficient.

6. For members of round tubing, use a length-to-diameter ratio of about 25:1.

7. Use overall structural shape to advantage. Arches, pyramids, folds, and steps span better than flat roofs. This is illustrated by the Biosphere II project (Fig. 8.9).

8. Consider the space truss-to-cladding interface when developing the design—the cleaner and simpler, the better.

COMPRESSION SYSTEMS

By following an ideal shape, arches and domes can resist uniform loads in compression, which is more efficient than flexure. Unbalanced live load and lateral load cause the line of action to

Figure 8.9 The helpful effects of dome, arched vault, and stepped pyramid shapes permit the long space truss spans seen here.
Biosphere II, Oracle, Ariz.; Architect: *Sarbid, Ltd.*; Space Frame Engineer: *Pearce Structures*. (Photo: *Peter Pearce, Pearce Structures*.)

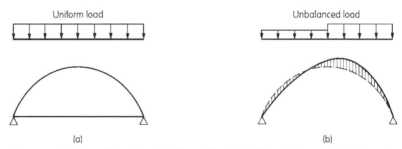

Figure 8.10 Arch considerations. (*a*) For uniform load, the arch line can be designed to follow the moment diagram, so only axial load occurs in ribs. (*b*) The moment diagram for unbalanced load deviates from the arch line, causing moment in ribs in addition to axial load.

deviate from an ideal shape, requiring some flexural strength and stiffness for stability (Fig. 8.10). This is generally provided by ribs. The ribs can be arranged so they do not meet. For a dome they would span radially from a central hub to a tension ring or thrust buttress at the spring line, and for an arched vault they would run perpendicular to the vault axis and parallel to each other (Fig. 8.11). Such an arrangement offers the advantage of straightforward rib fabrication and erection, but the disadvantage of large rib length and weight.

An alternative rib arrangement is a lamella pattern. For a dome, ribs curve out from the central hub in intersecting left- and right-hand spirals, like the patterns in the center of a sunflower or nested stars. For an arched vault, two sets of parallel ribs, each set skewed to the vault axis, intersect each other (Fig. 8.12). From an engineering viewpoint, the intersecting nature of lamella ribs offers excellent load redistribution ability. For erectors, an advantage is that the lamella ribs can be erected in short, self-supporting segments working inward from the dome spring line, like an Eskimo's igloo. For fabricators, lamella ribs pose a challenge as intersections are numerous and geometry is tricky. However, several proprietary systems are available, particularly for wood construction.

SHEAR/SHELL SYSTEMS

Warped surfaces such as hyperbolic paraboloids (hypars) offer the most efficient structural solution to spanning long distances, where the design can accept the rigid geometric requirements of the form (see Figs. 3.9 and 8.13). Hypar surfaces resist applied

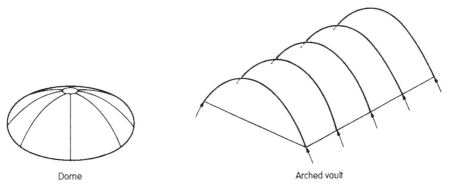

Dome   Arched vault

Figure 8.11 Conventional rib arrangements. (*a*) Dome (radial). (*b*) Arched vault (parallel).

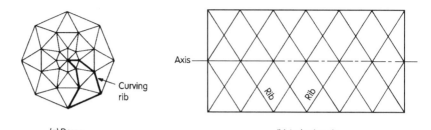

(a) Dome

Axis

Curving rib

Rib

Rib

(b) Arched vault

Figure 8.12 Lamella rib arrangements. (*a*) Dome has ribs in a spiraling crisscross pattern. (*b*) Arched vault has crossing ribs. (*c*) Arched steel vault with moment frame legs.
Kent College of Law, Illinois Institute of Technology, Chicago, Ill.; Architect and Engineer: *Holabird & Root.* (Photo: *McShane & Fleming Studios, courtesy of Zalk Joseph Fabricators Inc.*)

loads without the need for ribs, except for collectors at membrane edges.

While offering material efficiency, selection of a hypar must also consider construction methods and sound focusing. The curved surface of a hypar can be constructed entirely of straight elements, which is a convenience where concrete forms are made up of strips. However, mechanized placement and finishing of concrete is of limited value at the doubly curved surfaces here, so labor rates and

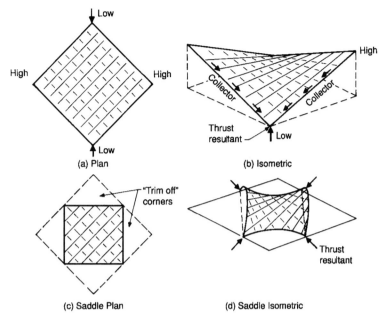

Figure 8.13 Hyperbolic paraboloids (hypars). (*a*) Plan. (*b*) Isometric. (*c*) Plan of saddle, formed by "trimming corners off" a hypar. (*d*) Saddle isometric.

scheduling must be considered when estimating the in-place structural cost.

An innovative hypar system used ribbed metal deck as the shear membrane element for cantilevering hypars at two hangars for jumbo jets (Fig. 8.14). A simpler, more common shear-based system is the folded-plate roof.

TENSILE STRUCTURES

In theory, a structure which uses members in tension should be much more efficient in material use than flexural or compression systems. As a practical matter, tensile structures have their place in many long-span applications, but they also have limitations. Such structures can be grouped as follows:

- Single- and double-layer radial cable nets
- Inverted arches
- "Tensegrity domes" or stacked rings
- Air-supported roofs
- Cable grids, nets, and tents

Figure 8.14 Hypars of metal decking are used here as the webs of cantilevers to provide column-free space and access all along the perimeter.
American Airlines Hangars, San Francisco and Los Angeles, Calif.; Architect and Engineer: *Conklin Rossant, Zetlin DeSimone and Chaplin JV.* (Photo: *Lev Zetlin Associates/Thornton-Tomasetti.*)

Radial Cable Nets

Radial cable nets require a central tension hub and an outer compression ring or series of buttresses. A single-layer system, such as New York's Madison Square Garden, uses a minimum of members, which speeds fabrication and erection. The cables and fittings used are similar in nature to those of relatively heavy bridge suspender rope. Stability of the roof against wind flutter and uneven loading is provided by the dead load of an intentionally heavy roof, which can add to the cost of the supporting structure. Also, drainage is at the center of this "inverted dome," so a fail-safe system against ponding is required (Fig. 8.15).

In a double-layer system upper and lower cables are connected to a common perimeter ring, but separate central hubs. By jacking the hubs apart the cable layers are prestressed against each other to

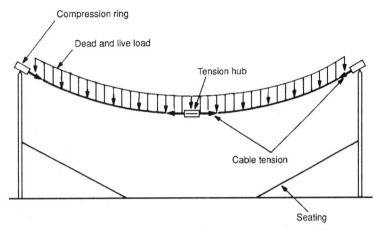

Figure 8.15 Section of a single-layer cable system showing tension hub, compression ring, and radial cables.

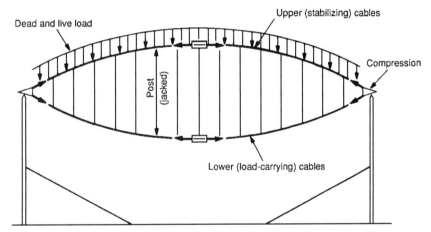

Figure 8.16 Section of a double-layer cable system showing a common compression ring and opposed cable sets with jacked struts between them.

provide shape stability without adding dead load (Fig. 8.16). The upper cables form a conventional dome shape for runoff to a perimeter gutter.

Inverted Arches

Inverted arches are appropriate for building roofs where headroom is not a factor. Here again dead load provides stabilization against wind flutter and unbalanced live loads. For long-span spaces the Dulles International Airport terminal in Washington, D.C., is a good example. The cables are buried within the roof slab, but the shape is the "exposed structure" (see Fig. 1.7). The Federal Reserve Bank in Minneapolis expresses an internal suspended arch in the facade treatment.

Stacked-Ring Systems

Stacked-ring and tensegrity dome systems are innovative recent developments. They use to full advantage current advances in fabricating, handling, and jacking cables similar to prestressing tendons. Bundles of light, strong, flexible tendons, easier to handle than a single wire rope of equivalent strength are sometimes used.

The deeper the sag in a cable, the less force is required in that cable to carry a given roof load. But as a practical matter, too great a sag will leave no headroom in the building. The solution was to break a single, long cable with generous sag into a series of cable segments and posts. The cables still have a slope matching the original, but the posts permit the segments to perch one above the other, preserving headroom. In concept it is similar to the Fresnel lens of a lighthouse, where the focusing effect of a thick lens is provided by a thin lens with rings of sloped glass (Fig. 8.17).

In stacked-ring and tensegrity systems a perimeter compression ring is needed. In addition tension hoops are required at each ring and posts are required to spread the cables in a complex web.

These domes come in two configurations. The stacked-ring system is more planar, with chord and web cables and posts aligned in vertical planes extending radially from the central hub. This simplifies analysis, design, and construction, but requires great care in maintaining structural stability as large deflections can occur before the forces to restrain the vertical posts against rollover are generated. The tensegrity dome system is more three-dimensional, as tendons are arranged in a crisscross pattern (see Figs. 3.22 and 3.23). This adds to overall stability and the ability to redistribute local loads, but makes jacking more complicated.

Advantages of these systems include ease in supplying and handling the light cable, provision of a domed-shape roof profile for good drainage, and, perhaps most important, the ability to assemble a system at grade and erect it by just jacking some of the cables. The light weight of the structural components helps make them economical. The primary disadvantage is the need for the engineers, contractors, and erectors to be on top of the design and construction at all times, as the systems are new, sensitive, and complex.

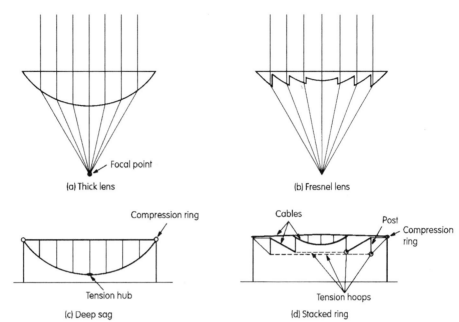

Figure 8.17 Analogy between a stacked-ring or tensegrity cable system and a Fresnel lens—both depend on slopes for effectiveness. (*a*) A conventional lens can be quite thick to get the desired convexity. (*b*) A Fresnel lens keeps the convex slope but removes unnecessary material. (*c*) A conventional cable system with an efficient sag-to-span ratio requires excessive headroom. (*d*) A stacked-ring system uses posts and hoops to maintain effective slopes within minimal depth.

Air-Supported Roofs

Air-supported roofs can provide a conventional domed shape for good drainage, but qualify as tensile structures due to the tension in fabric and hold-down cables to resist internal air pressure. In terms of exposed structure, there is very little to see—some fabric, some cables, and sometimes draft curtains. In this system support comes from the mechanical equipment room.

While air-supported structures have been heavily promoted and widely used, they are not universally applicable. First, to control air losses, entrances must be limited and of a low-leakage type—vestibules or revolving doors. Second, conventional construction is still required for sidewalls all around the enclosed space, including a compression ring around the perimeter. Third, it is impractical to maintain heavy and unbalanced live loads with air pressure alone, which may lead to sagging and leaking at some snowy sites. Fourth, wind flutter can be a concern given the flat profile and negligible

Figure 8.18 This fabric roof by Birdair, Inc., spanning 90 m (300 ft), uses two way curvature to provide rigidity against destructive flapping.
San Diego Convention Center, San Diego, Calif.; Architect: *Arthur Erikson*; Engineer: *Horst Berger Partners*. (Photo: *Birdair Inc.*)

stiffness of typical roofs. And fifth, fabric roofs do not have the life expectancy of conventional construction. A low first cost may get a stadium covered, but will the owners have the funds to replace the cover when necessary? Replacement costs will be considerably higher than for conventional reroofing.

Cable Grids and Tents

Cable grids and tents look particularly simple and straightforward to the public eye, but can be the most complex of all systems to design and construct.

The main challenge in cable grids and tents is to avoid destructive flutter. Since the materials used do not have the flexural stiffness of the ribs of a dome or arch, stiffness must be provided by geometry. Typically this takes the form of reverse curvatures, where one set of cables crosses another cable set or tent fabric curving in the opposite direction (Fig. 8.18). Geometry is the structure, and this can restrict the architect from making major changes once basic parameters are set. Also, a series of poles and struts will be required at various points.

Cables must be cut to exact length or jacked to create the desired prestress effect. Tent fabric must be cut and sewn on a bias to get the required three-dimensional shape—a talent known to sailmakers but not most builders. Even after prestressing, cable nets deflect much more under unbalanced load than other, stiffer systems. Such conditions should be investigated using nonlinear or incremental analysis programs, which are unusual for building structural design.

Finally, the tent materials used to date have significantly shorter projected lifespans than that usually assumed by owners and the public. We are accustomed to think of 20-year reroofing of a dome as normal maintenance, but recovering a tent in new fabric somehow seems more drastic. It certainly requires more specialized contractors.

CONCLUSION

Long-span spaces have always captured the imagination of designers, builders, and the public. Exposed or expressed structural systems are typically part of these projects, adding excitement, scale, and texture. With modern materials and methods of analysis the structural options available are greater than ever. The biggest challenge now is to determine which option is most suitable for a particular project.

BIBLIOGRAPHY

ACI Committee 303 (rev. 1982): "Guide to Cast in Place Architectural Concrete Practice," American Concrete Institute.

ACI 318-89 (1989): *Building Code Requirements for Reinforced Concrete* (and 318 R-89: *Commentary*), American Concrete Institute, Detroit, Mich.

ACI Committee 350 (May–June 1989): "Environmental Engineering Concrete Structures," *ACI Structural Journal.*

AITC 104-84 (1984): "Typical Construction Details," American Institute of Timber Construction.

AITC 109-90 (1990): "Standard for Preservative Treatment of Structural Glued Laminated Timber," American Institute of Timber Construction.

AITC Technical Note 11 (rev. 1987): "Checking in Glued Laminated Timber," American Institute of Timber Construction.

AITC Technical Note 12 (November 1986): "Designing Structural Glued Laminated Timber for Permanence," American Institute of Timber Construction.

Ammar, R., Carino, N., and Fowler, D. W. (1975): "Use of Sulfur to Repair Damaged Concrete," in *Polymers in Concrete*, American Concrete Institute.

Architectural Precast Concrete, 2d ed. (1989), Precast/Prestressed Concrete Institute.

"Arena Roof Design Saves Views, Budget" (June 9, 1977), *Engineering News-Record*, McGraw-Hill [Rutgers University Athletic facility.]

ASHRAE *Handbook: Fundamentals* (1989), American Society of Heating, Refrigerating, and Air Conditioning Engineers, Inc., Atlanta, Ga.

ASTM Standard Specifications A143, C33, C227, C289, C666, E547 (updated irregularly), American Society for Testing and Materials, Philadelphia, Pa.

Berke, N. S., Pfeifer, D. W., and Weil, T. G. (December 1988): "Protection against Chloride-Induced Corrosion," *Concrete International.*

Brannigan, F. (1982): *Building Construction for the Fire Service*, 2d ed., National Fire Protection.

Brevoort, G., and A. Roebuck, (February 1991): "Selecting Cost-Effective Protective Coating Systems," *Materials Performance Magazine.* [Compares coating performance and relative life-cycle costs.]

Brockenbrough, R.L., and Johnston, B.G. (1981): *Steel Design Manual,* U.S. Steel Corp., Pittsburgh, Pa., chap. 1.

Brown, S., and Usui, N. (September 8, 1988): "Shortcuts to Long Spans," *Engineering News-Record.* [New designs increase use of off-the-shelf structural systems.]

Burns, M., and Rizzo, E. M. (December 1989): "The Durability of Exposed Aggregate Concrete," *Concrete International.*

Cassera, D., and Feist, W., (1991): "Exterior Wood in the South," USDA Forest Products Lab. GTR-69. [Recommendations on details, finish treatments, durability.]

Chrest, A. P. (November 1988): "Designing Concrete Parking Structures for Long Term Durability," *Concrete International.*

Cohen, M., and Olek, J. (November 1989): "Silica Fume in PCC: The Effects of Form on Engineering Performance," *Concrete International.*

Color and Texture in Architectural Concrete (1980), Portland Cement Association.

"Concrete Degradation due to Thermal Incompatibility of Its Components" (1989–1990), *Journal of Materials in Civil Engineering.*

"Concrete Inspection Guidelines" (1975), Portland Cement Association. [The color of concrete.]

Cuoco, D. A. (1985): "Design and Construction of Space Frame Structures," 1985 International Engineering Symposium on Structural Steel.

Cuoco, D. A. (June 1982): "Today's Space Frame Structures: Sophisticated, Adaptable, Reliable," *Architectural Record.*

Dannemann, R. (1989): "Discussion of 'Modification of Behavior of Double-Layer Grids: Overview,' " *Journal of the Structural Division,* ASCE, p. 1570.

"Design for Fire Resistance of Precast Prestressed Concrete" (1989), Prestressed Concrete Institute.

"Design Guide and Commentary to Wood Structures" (1975), American Society of Civil Engineers.

"Development of Fire Resistant Steel (FR Steel) for Buildings," Nippon Steel Corp.

Dietsch, D. (May 1985): "High Tech Expansion," *Architectural Record.* [O'Hare Airport.]

Dunstan, M. R. H. (1986): "Fly Ash as the Fourth Ingredient in Concrete Mixtures," in *Fly Ash, Silica Fume, Slag and Natural Pozzolans in Concrete*, vol. 1, American Concrete Institute.

Durning, T. A., and Hicks, M. C. (March 1991): "Using Microsilica to Increase Concrete's Resistance to Aggressive Chemicals," *Concrete International.*

Eglinton, M. S. (1987): "Concrete and Its Chemical Behavior," American Society of Civil Engineers.

Elnimeiri, M. (1986): "Onterie Center," in *Advances in Tall Buildings*, Council on Tall Buildings and Urban Habitats, Van Nostrand Reinhold.

Emmerich, R. W.: "Decay in Wood Structures," in *Wood Structures Decay.*

"Engineers Exploit Spatial Geometries for Economy and Architectural Fitness" (August 1975), *Architectural Record.*

"Evaluation, Upgrading and Maintenance of Wood Structures" (1982), American Society of Civil Engineers.

Fire Safe Structural Steel—A Design Guide, American Iron and Steel Institute.

"Fire Safety of External Building Elements," *AISC Engineering Journal*, American Institute of Steel Construction.

Fletcher., Sir B. (1987): *A History of Architecture*, Butterworth & Co, London.

Foster, N. (1983): "Headquarters for the Hong Kong and Shanghai Banking Corporation," in *Developments in Tall Buildings*, Council on Tall Buildings and Urban Habitats, Van Nostrand Reinhold.

Fowler, D. W.: "Innovative Uses of Polymer Concrete," in *Serviceability and Durability of Construction Materials*, vol. 1.

Gaidis, J. M., and Rosenberg, A. M. (April 1989): "A New Mineral Admixture for High Strength Concrete," *Concrete International.*

Galuszka, P. (June 19, 1986): "Atriums Roll with Chicago Tower," *Engineering News-Record.*

Green, P. (February 2, 1989): "Stockholm Scales Up Sphere, Largest Spherical Building in the World," *Engineering News-Record.*

Gregerson, J. (August 1988): "Chicago's Library Competition Yields a Neoclassic," *Building Design and Construction.*

Griffin, D. F.(1975): "Corrosion Inhibitors for Reinforced Concrete," in *Corrosion of Metals in Concrete,* American Concrete Institute.

Gustafson, K. (April 1986): "INTELSAT Launches Its World HQ," *Corporate Design and Realty.*

Gutman, A., and Tomasetti, R. L. (October 27, 1986): "Battery Park Centerpiece Uses Steel Extensively," *New York Construction News.*

Hanaor, A., March, C., and Parke, G. (May 1989): "Modification of Behavior of Double Layer Grids: Overview," *Journal of the Structural Division, ASCE,* vol. 115.

IABSE Structures C-23/82 (1982): *Selected Works of Fazlur R. Khan,* International Association for Bridge and Structural Engineering.

Iyengar, H.: "Exposed Steel Frame—A Unique Solution for Broadgate, London," draft.

Korman, R. (October 9, 1986): "A Terminal Framed with Romance," *Engineering News-Record.* [O'Hare Airport.]

Kurz, G. (May 1990): "Dust Free Pigments Make the Coloring of Concrete a Pleasure," *Concrete Plant and Production.*

Larrabee, C. P. (August 1953): "Corrosion Resistance of High-Strength Low-Alloy Steels as Influenced by Composition and Environment," *Corrosion Magazine,* vol. 9, 8., National Association of Corrosion Engineers.

Litvin, A.: "Clear Coatings for Exposed Architectural Concrete," Bulletin D137, Portland Cement Association.

"Marathon Project Tops Touchy Site" (October 14, 1982), *Engineering News-Record.* [Copley Place project.]

Makowski, Z. S. ed., (1981) *Analysis, Design and Construction of Double-Layer Grids,* Halsted Press, New York, New York.

Marusin, S. (1987): "Improvement of Concrete Durability against Intrusion of Chloride-Laden Water by Using Sealers, Coatings and Various Admixtures," in *Concrete Durability: Katherine and Bryant Mather International Conference,* vol. 1.

Marusin, S.: "Enhancing Concrete Durability by Treatment with Sealers," *Structural Materials Journal.*

Metals Handbook, Boyer and Gail, eds. (1985) p.4–85, ASM International, Materials Park, Ohio.

"One Mellon Bank Center, Pittsburgh, Pa.," U.S. Steel Building Report (1984).

Ong, K., and Ravindrarajah (1987): "Corrosion of Steel in Concrete in Relation to Bar Diameter and Cover Thickness," in *Concrete Durability: Katherine and Bryant Mather International Conference*, vol. 2.

"Parking Structures: Recommended Practice for Design and Construction" (1988), Prestressed Concrete Institute.

Pei, I. M., and Partners: "Responsibility for Formwork and Approvals," memorandum.

Pickard, S. S. (May 1991): "Construction Oriented Concrete Design," *Concrete International*.

Robinson, R. C. (1975): "Cathodic Protection of Steel in Concrete," in *Corrosion of Metals in Concrete*, American Concrete Institute.

Robison, R. (December 1984): "High Tech Is More than a Look," *Civil Engineering/ASCE*. [Tubular steel structures.]

Robison, R. (October 1988): "Weathering Steel: Industry's Stepchild," *Civil Engineering/ASCE*.

Rutes, W. A. (October 1968): "A New Look at Office Buildings," *Architectural and Engineering News*.

Scharfe, T. (March 1988): "Focus on Glass," *Building Design and Construction*.

Schupack, M. (February 1991): "Corrosion Protection for Unbonded Tendons," *Concrete International*.

"Slip-Forming/Flying Forms," Bulletin 24, *Concrete Reinforcing Steel Institute*.

"Spaceframes—The Solution." (1991), MERO Structures, Inc., promotional literature.

"Ten Rules of Thumb for Space Frame Design" (1988), Pierce Structures, promotional literature.

"The Annual of Technology" (November 1986): *Building Design & Construction*. [State-of-the-art applications of products and systems.]

Thornton, C. H. (1986): "Northwest Atrium—Steel Adapts to Complex Geometry," *Modern Steel Construction*, 3d quarter.

Thornton, C. H., and Tomasetti, R. L. (November 1970): "Hangar Features Stressed-Skin Hypars," *Civil Engineering/ASCE*.

TnemecTopics, vols. 2–9, Tnemec Co., Inc., Kansas City, Mo., promotional literature. [Discussions of coating systems.]

Tomasetti, R. L. (September 1972): "New Approaches to the Fire Protection of Steel," *Architectural Record.*

Tomasetti, R. L., et al. (1986): "A Stressed Skin Tube Tower: One Mellon Bank Center," in *Advances in Tall Buildings*, Council on Tall Buildings and Urban Habitats, Van Nostrand Reinhold.

Townsend, C. L. (1968): "Control of Temperature Cracking in Mass Concrete," in *Causes, Mechanism and Control of Cracking in Concrete*, American Concrete Institute.

"Two Way King Post Truss Carries Lightweight Auditorium Roof" (October 23, 1975), *Engineering News-Record.*

Villecco, M. (January/February 1971): "Giant Steel Hangar's Cantilevered Roofs Will Shelter Four Jumbo Planes," *Architectural Forum.*

Wall, D. (October 27, 1986): "Steel Allows Unusual Shapes for IBM," *New York Construction News.*

White, R., and Salmon, C. Eds (1987): *Building Structural Design Handbook*, Wiley, Chap. 10.

Wood, J. S.: "The Nature of Concrete," memorandum.

Wood, J. S.: "Office Buildings—Definitions, Descriptions and History," in *Wilkes Encyclopedia of Architecture*, vol. 3, Van Nostrand Reinhold.

Wood, J. S.: "Liberty Center, Pittsburgh, Pa.," draft.

Wood, J. S.: "Westin Hotel, Copley Place, Boston, MA," draft text and fact sheet.

Wood, J. S. (1988): "Prescott Mill Housing—Reuse of a New England Heritage," in *Wilkes Encyclopedia of Architecture*, Van Nostrand Reinhold.

Wright, G. (March 1988): "Curtain Walls Drawing More Critical Attention," *Building Design and Construction.*

Zunz, J., and Glover, M. (1986) "The Hong Kong and Shanghai Bank Projects," in *Advances in Tall Buildings*, Council on Tall Buildings and Urban Habitats, Van Nostrand Reinhold.

Index
